THE BABY FOOD DIET

A Modern Lifestyle Philosophy

John Probandt

婴儿食品节食法

一种现代生活方式理念

[美] 江柏德 —— 著
王梅 —— 译

华夏出版社
HUAXIA PUBLISHING HOUSE

图书在版编目（CIP）数据

婴儿食品节食法：一种现代生活方式理念/（美）江柏德著.--北京：华夏出版社有限公司，2019.11
ISBN 978-7-5080-9865-4

Ⅰ.①婴… Ⅱ.①江… Ⅲ.①婴儿－减肥－食谱 Ⅳ.①TS972.161

中国版本图书馆CIP数据核字(2019)第218729号

婴儿食品节食法——一种现代生活方式理念

著　　者	[美]江柏德
译　　者	王　梅
责任编辑	梅　子　阿　修
出版发行	华夏出版社有限公司
经　　销	新华书店
印　　装	三河市万龙印装有限公司
版　　次	2019年11月北京第1版 2019年11月北京第1次印刷
开　　本	880×1230　1/32开
印　　张	9.25
字　　数	220千字
定　　价	45.00元

华夏出版社有限公司　地址：北京市东直门外香河园北里4号
邮编：100028　网址：www.hxph.com.cn
电话：(010) 64663331（转）

若发现本版图书有印装质量问题，请与我社营销中心联系调换。

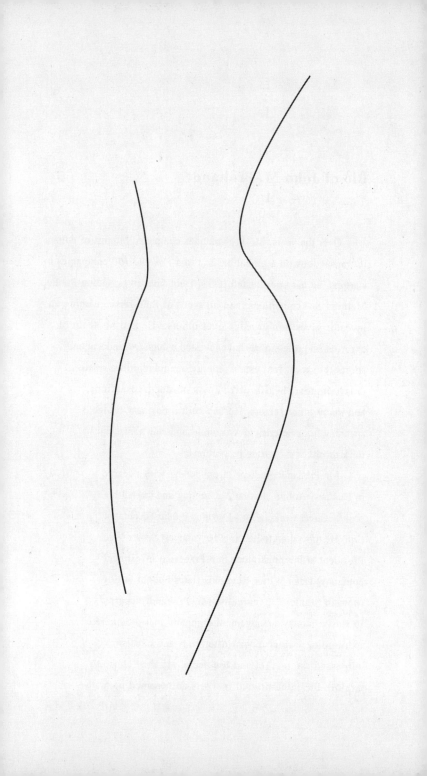

Bio of John M. Probandt

Over the years, Mr. Probandt has completed billions of dollars in transactions on a global basis. From *Fortune 100* companies to startups, he has coordinated IPO's, Debt financings, Going Public, Mergers & Acquisitions, and all types of fund raising. Being an investor gives him an edge over others. He and his team of experienced professionals have invested in high tech, education, sports, bio tech, real estate, energy, manufacturing, retail, entertainment, health care, cross border transactions, and many other sectors. He has built a business bridge spanning from America to Asia while amassing a financial track record of significant proportions.

Mr. Probandt graduated from college with a degree in Business Administration/Accounting and started his prolific career working for a Certified Public Accountant firm. He moved on to become the youngest Senior Vice President at the prestigious San Francisco investment company, Dean Witter Reynolds (later bought out by Morgan Stanley). Eventually, Mr. Probandt departed to start a money management company that produced extraordinary returns year after year and consistently surpassed the market and his peers. He then chose to explore the international markets and teamed up with

江柏德先生小传

多年来,江柏德先生(Mr. Probandt)在全球范围内已完成了数十亿美元的交易。从**财富一百强公司**到新创业公司,他先后经手协调过:首次公开发行新股、债券金融、公司上市及企业并购等各种资金筹措,相较其他人士,作为一名投资人他更具有优势。他率领其拥有资深人士的团队在诸多领域做过投资,涉及高科技、教育、体育、生物科技、房地产、能源、制造、零售、娱乐、医疗保健、跨境电商交易以及其他许多领域,他建立了一座横跨美洲到亚洲的商业之桥,并在金融方面取得了重大成就。

大学时代,江柏德先生主修工商管理和会计学,他丰富的工作生涯始于一家注册会计师事务所,之后他跳槽到旧金山一家著名的投资公司——添惠证券投资(Dean Witter Reynolds,后被摩根士坦利投资公司 Morgan Stanley 收购)出任最年轻的高级副总裁。最终,江柏德先生离开前公司,成立了一家资金管理公司,年复一年,创造了非凡的收益,大大超越市场均值与同行业绩。进而,他选择开拓国际市场,并

a Chinese enterprise, The CITIC Group, and moved to Beijing, China, with the initial project of bringing Marvel Entertainment to Asia (film, merchandising, tourism developments).

Mr. Probandt has over thirty (30) years of experience, investing in the fitness, health, nutrition, diet and bioscience industries. Through various funds he has been a member, responsible for the completion with investing of over three hundred million USD (300,000,000.00), in privately held enterprises. In addition, he has been responsible for the deployment of well over two and a half billion dollars USD (2,500,000,000.00) in publicly traded health sector opportunities. Mr. Probandt has been an owner, investor and innovator within health, nutrition, diet and fitness industry for over three decades.

At one point, Mr. Probandt owned a chain of gyms, "Imperial Fitness". The clubs primarily focused on exercise, dieting and supplements. At this time Mr. Probandt realized that relying on membership fees alone was not enough to enhance the profitability of the fitness chain. He did two things to address this need: 1) Launch a line of vitamin supplements. He was involved in every phase of the development and 2) Mr. Probandt's clubs were among the first to offer members personal trainers in America. It is now normal in every club in the world, but at the time it was a novel idea.

Through another fund, of which Mr. Probandt was

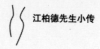 江柏德先生小传

与中国企业中信集团（CITIC）合作，他本人也因此移居中国北京。他们最初的合作项目，是将漫威娱乐带到亚洲（电影、营销、旅游开发）。

江柏德先生在健身、健康、营养、节食和生物科学等领域的投资经验超过三十年。通过其为会员的各项基金，负责完成了对民营企业投资，总额超过3亿美元。此外，他还负责将超过25亿美元的资金用于卫生领域的上市投资。三十多年来，江柏德先生一直是健康、营养、节食和健身行业的参与者、投资者和创新者。

一度，江柏德先生曾经拥有"帝国健身"俱乐部连锁店，这些俱乐部主要专注于锻炼、节食和营养补充剂。当时，江柏德先生已经意识到，仅仅依靠会员费不足以提高健身连锁店的收益率。为了满足这一需求，他做了两件事：1）推出系列维生素营养补充剂，并且他本人亲自参与到其发展的每一阶段；2）江柏德先生的健身俱乐部是美国最早向会员提供私人教练服务的俱乐部之一。现在，这一理念在世界上的各个俱乐部都很常见，但在当时还是非常新颖、前卫的。

a partner a large proportion was acquired in a publicly traded supplement company. The original idea was to acquire controlling stakes in the company, and then to expand it to other parts of the world, including China. That brand did expand in China and other places due to Mr. Probandt's influence, but the proportion was eventually sold due to conflicts with top management.

Mr. Probandt understands the diet, nutrition and fitness industry, as it is one of his hobbies and passions. He is an avid reader on the subject and in the past has been a speaker at many health conferences.

Mr. Probandt is the co-founder of the Private Equity (PE) enterprise, *Checkmate Investing,* which continues its legacy of making and sustaining successful investments. Mr. Probandt believes in teaming up with larger partners to assure pipeline and success. The Checkmate team is comprised of top tier experts in the financial world. Mr. Probandt's book Checkmate Investing is published by China Law Press. Checkmate has a silo "verticals" approach to investing with funds in: Entertainment Tech/Immersive Entertainment, Biotech, Tourism Real Estate, Agriculture Tech and Energy.

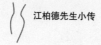 江柏德先生小传

江柏德先生通过其作为合伙人的另一只基金,在一家营养补充剂上市公司中获得了大量头寸,他最初的想法是获得该公司的控股权,然后将业务扩展到全球其他国家和地区,比如中国。在江柏德先生的影响之下,该品牌的确在中国及其他区域进行了扩张,但由于其与最高管理层的分歧,头寸最终被出售。

江柏德先生洞悉节食、营养和健身行业,因为这正是其爱好和激情所在。他是这一领域的热心读者,也曾是许多健康会议上的发言人。

目前,江柏德先生出任一家私募股权(PE)企业——**将军投资**(Checkmate Investing)的联合创始人,并延续了其创造和保持的成功投资的传统。江柏德先生坚信要与更大型的企业合作,以确保合作顺畅和成功。**将军投资**团队由金融界的顶级专家组成。江柏德先生所著《一击制胜》(*Checkmate Investing*)一书已由中国法律出版社出版。**将军投资**提出了竖井垂直法来进行投资,涉足娱乐科技/沉浸式娱乐、生物科技、旅游房地产、农业科技和能源等领域。

Recommendation

The Baby Food Diet is a novel way for dieters everywhere to implement portion control. Simply, a great way to lose and keep weight off.

——Jeffrey Hsu MD, Cardiologist and former
American Council on Exercise, personal
trainer and certified instructor

推荐语

《婴儿食品节食法》是为世界各地的节食者进行饮食分量控制的一种新奇方法。简单来说,这是一个很棒的减肥及保持体重的方法。

——杰弗里·许(Jeffrey Hsu)
医学博士,心脏病学家,
曾任职于美国运动协会,
私人教练和认证教练

My name is Tony Villa, and I am thrilled to write the foreword to this book, *The Baby Food Diet*. I have been directly involved in the ideas presented herein and can assure that this is the only diet in the world that never stops giving benefits and will change your life for the better. It is not based on science, but instead on the old fashion idea of observation and logic.

I am a businessman and professional singer. The difference between a good song and a great song transpires when the singer *becomes* the song. The same can be said about this diet: it will become you, and a new person within you will emerge.

We would like to refer to *The Baby Food Diet* as *The Forever Diet* as it will allow you to reach your desired weight and keep you there forever while improving both your mental outlook and physical health.

———Tony Villa

推荐语

我叫托尼·维拉（Tony Villa），非常激动能为本书——《婴儿食品节食法》作推荐。我直接参与提炼书中所呈现的理念，并可以保证，这是世界上唯一一种能够不断提供益处，并使你的生活变得更好的节食方法。它不是建立在科学的基础上，而是基于观察和逻辑。

我是一名商人及职业歌手。一首不错的歌曲和一首非常棒的歌曲之间的区别会消失，如果歌手将自身融入歌曲之中。同样的道理也适用于这种节食方法：它会融入你，从你的内心中呈现出一个全新的自己。

我们喜欢将**婴儿食品节食法**称为永久节食法，因为它能让你达到理想体重，并且长久保持，同时还能改善精神面貌和身体健康。

——托尼·维拉（Tony Villa）

contents

CHAPTER 1 The Golden Solution to Weight Loss xiv

CHAPTER 2 Three Variations of the Diet 12

CHAPTER 3 Knowing Your Weight 20

CHAPTER 4 Exercise and Health 34

CHAPTER 5 Mindset and Other Interesting Thoughts 54

CHAPTER 6 Benefits of The Baby Food Diet 64

CHAPTER 7 Protein and Baby Food 78

CHAPTER 8 Psychological Part of the Diet 92

CHAPTER 9 Exercise 102

CHAPTER 10 Losing Weight Requires Commitment 118

CHAPTER 11 Body, Life and Balance 126

CHAPTER 12 Alcohol and Supplements 140

CHAPTER 13 100 Amazing Suggestions 162

CHAPTER 14 Short Summary of the Baby Food Diet 238

ACKNOWLEDGMENTS 248

BONUS CHAPTER A Few More Thoughts 258

目录

第一章　黄金减肥方案　1

第二章　节食的三个变化阶段　13

第三章　自知体重　21

第四章　运动与健康　35

第五章　心态和其他有趣的想法　55

第六章　婴儿食品节食法的益处　65

第七章　蛋白质和婴儿食品　79

第八章　节食中的心理因素　93

第九章　运动　103

第十章　减肥需要承诺　119

第十一章　身体、生活和平衡　127

第十二章　酒精和营养补充剂　141

第十三章　100个绝佳的建议　163

第十四章　婴儿食品节食法概要　239

致谢　249

附录　再说几点　259

CHAPTER 1

The Golden Solution to Weight Loss

第一章

黄金减肥方案

I will start this book off with a very bold statement: *The Baby Food Diet* is a perfect diet. I say that because it will change your life in such a positive, exciting way that you will soon be telling everyone about this permanent solution to weight management and a happy life. This easy-to-follow diet will improve both your physical and mental balance and transform you into a superstar in your own eyes and the eyes of others. This is a fact, not an exaggeration. *The Baby Food Diet* will help you lose weight, reach your desired proportion, and maintain that size forever.

Let me provide a brief summation of how my weight-loss journey lead to *The Baby Food Diet*. It is important to understand that many, many factors contribute to gaining or losing body fat. You must assess every aspect of your past and present life, then decide who and where you want to be in the future. *The Baby Food Diet* is more than a guide to healthy consumption. It is also a practical lifestyle philosophy that will benefit everyone.

When I was growing up, I watched my mother struggle with her weight. She was an attractive woman and by no means obese. The small amount of extra weight she carried was always on her mind. She tried numerous ways to shed these few extra pounds, but none offered a long-term solution. I wish she could have had *The Baby Food Diet*, a permanent solution to getting the weight off and keeping it off. By the way, her overweight

第一章
黄金减肥方案

我将以一个非常大胆的声明作为本书的开始：**婴儿食品节食法**是完美的节食方法。之所以这么说，是因为它将以十分积极的、令人兴奋的方式改变你的生活，你很快就会把这种永久的减重和幸福生活的方法告诉所有人。这种简单易学的节食方法，可以改善你的身心平衡，让你在自己和别人眼中都成为超级巨星。这是事实，并非夸张。**婴儿食品节食法**将帮助你减肥，达到你想要的身材比例，并永久保持体形。

请容我简要总结一下我自身的减肥之旅，是如何将我引向**婴儿食品节食法**的。许许多多因素都会有助于增加或减少身体脂肪，理解这一点很重要。你必须评估自己过去和现在生活的方方面面，然后决定自己将来想要成为什么样的人，以及走什么样的路。**婴儿食品节食法**不仅仅是健康饮食的指南，也是一种实用的生活哲学，会让每个人从中受益。

在我成长的过程中，曾目睹我的母亲怎样与自身体重做斗争。她是个迷人的女人，根本算不上肥胖，但她身上那一点点额外的重量却一直是她的心病。为了减掉这几磅赘肉，她尝试了很多方法，但没有一种能够提供长期的解决方案。我真希望她可以尝试**婴儿食品节食法**——这种

issue was the only trait of my wonderful mother that I didn't want to inherit.

With her in mind, my goal is to help others with the knowledge I have discovered through years of failures and successes. It is now clear to me that being thin and staying thin does not need to be a lifelong battle. I want everyone to know that *The Baby Food Diet* can help them realize their new normal, a thin and healthy body.

Before trying *The Baby Food Diet*, I, like my mother, tried a plethora of diets. With each one, I would lose weight and feel happy about myself. Then after a while, I would always regain the original weight, plus more! Those other diets offered no long-term solutions. With *The Baby Food Diet*, I have always stayed thin.

Now let me explain exactly how *The Baby Food Diet* came into being.

Many years ago, while awaiting the departure of my aircraft from Los Angeles to Zurich, Switzerland, I sat patiently in my assigned seat. Soon, a very attractive and shapely flight attendant asked what I would like to drink. She also inquired if I would like to read a newspaper and if so which one. When I told her my preference, she informed me the airline did not carry that brand. Just as we were about to depart, she returned with my requested newspaper. Overjoyed, I questioned where she had gotten it. She had run off the plane and bought it

第一章
黄金减肥方案

永久减肥并保持体形的方法。顺便说一下,超重问题是我唯一不想从了不起的母亲那里继承的特点。

由于她的缘故,我的目标就是用自己从多年的失败和成功的经验教训中所总结的知识、方法去帮助他人实现减重。现在我很清楚,一个好的身材和长久的保持并不是一场终身的战斗。我想让每个人都知道,**婴儿食品节食法**可以帮助他们建立新常态,做一个苗条又健康的人。

在尝试**婴儿食品节食法**之前,我像母亲一样,尝试过许多不同的节食方法。每次尝试,我都会减轻体重,我为自己而感到高兴。然而过了一段时间之后,我总是又回到了原来的体重,甚至还更重!那些节食方法都不能提供长期的解决方案,然而运用**婴儿食品节食法**,却总能令我保持体形。

现在让我来详细讲述一下**婴儿食品节食法**是如何诞生的。

多年以前,在从洛杉矶飞往瑞士苏黎世的航班上等待起飞时,我耐心地坐在我的座位上。不久,一位非常迷人、身材匀称的空姐问我想喝什么饮料,她还问我是

in the airport lobby. Remember, this was years ago. With today's strict rules, it is doubtful she would even be allowed to exit the plane to buy a newspaper. Fact is, in this digital world, very few airlines provide newspapers anymore. Needless to say, I was very appreciative. Since the plane was almost empty, we spent much of the flight talking and became good friends.

When it came time for dinner, she invited me to join her in the back. Due to the light passenger load, she and other flight attendants had little to do. The meal was rather magnificent which makes this story even more unbelievable, as most airlines serve less than delicious food. As I started to eat, she sat down beside me and pulled out two jars of baby food. That's right, two jars of pureed baby food! I curiously inquired about her meal choice, and she merely asked if I liked her figure. I told her without question that she was absolutely beautiful, especially when it came to her perfect physique. She went on to divulge her secret, which was to substitute two meals each day with baby food. She enthusiastically quantified that baby food takes weight off and keeps it off. We remained friends for several years and often joked about *The Baby Food Diet* that contributed to her lasting beauty.

In the ensuing years, I was on a weight "roller-coaster", gaining and losing pounds by trying various dieting methods. While I was successful enough each time,

第一章
黄金减肥方案

否想看报纸,如果想看,喜欢哪种报纸。当我告诉她想看的报纸时,她说这家航空公司没有那种报纸。正当飞机要起飞的时候,她拿着我要的报纸回来了。我喜出望外,问她从哪儿弄来的。她是从飞机上跑下去,在候机大厅买到的。请记住,这是很多年前的事了。按如今的严格规定,她根本就不能下飞机去买报纸。事实上,在如今的数字世界里,已经很少有航空公司提供报纸了。自不必说,我万分感激。由于飞机上几乎没什么乘客,在飞行中我们基本上一直都在聊天,并成了好朋友。

到了用晚餐的时间,她邀请我一起坐在机舱后部。由于乘客不多,她和其他空乘人员没太多事可做。这顿饭相当丰盛,使得这个故事变得更加令人难以置信。因为大多数航空公司提供的餐食都不太好吃,当我开始进餐的时候,她坐在我旁边,拿出了两罐婴儿食品。没错,是两罐婴儿食品泥!我好奇地询问她为什么选择吃这个,她只是问我是否喜欢她的身材。我毫无疑问地告诉她:非常漂亮,尤其是她那完美的体形。接着她透露了个中奥秘,那就是每天用婴儿食品代替两顿饭。她热心地证实了婴儿食品可以减肥并且不会反弹。我们保持了好多年的朋友关系,经常会拿**婴儿食品节食法**开玩笑,正是这

the problem was that the weight always came back. Each time, I would end up heavier and unhealthier than before. Then one day, I was walking along the street and for some reason thoughts of this pretty flight attendant came to mind. I remembered her baby food diet and decided it was time to try her approach to not only losing weight but keeping it off.

That night, I bought a big supply of baby food and substituted two jars per two meals each day. For the third meal, I decided to eat something high in protein. Somehow, I just knew it was integral to stay away from too many carbohydrates beyond what the baby food provided. This is not a Keto diet as baby food itself is rich in natural carbohydrates. However, these are natural carbohydrates, not processed carbohydrates from foods such as desserts, bread and noodles.

I expected the baby food to taste awful. In fact, it took me one week after buying all the baby food to drum up the courage to start the diet. When at last I got the nerve to take the first bite, I was happily surprised; the baby food tasted excellent! It was rich in nutrients and very healthy. Baby food is pureed, so it is easy to eat and digest. After about eight weeks on the diet, the weight was gone. I was a new man. Even better, this time the weight stayed off! Thankfully, *The Baby Food Diet* was born.

Like all diets, the first two weeks were trying.

第一章
黄金减肥方案

种方法让她的美丽经久不衰。

在接下来的几年里,我经历了体重的"过山车",通过尝试各种节食方法,来回增重和减重。虽然我每次都很成功,但问题是体重总会反弹。每次尝试的结果都是,我会比以前更胖,更不健康。直到有一天,我走在街上时,不知何故,突然想起了这位漂亮的空姐,想起了她的婴儿食品节食法,我决定是时候试试她的方法了,不仅要减重,而且要持久。

那天晚上,我买了大量的婴儿食品,每天用其中的两罐替换两餐。第三顿饭,我决定吃一些高蛋白的食物,我明白除了婴儿食品所提供的之外,碳水化合物能少则少的必要性。这不是生酮饮食法,因为婴儿食品本身含有丰富的天然碳水化合物。当然,这些都是天然的碳水化合物,而不是来自甜点、面包和面条等食物的加工过的碳水化合物。

我原以为婴儿食品会很难吃,事实上,在把所有婴儿食品都买回来以后,我花了一个星期的时间才鼓起勇气开始尝试节食。当我终于鼓起勇气吃了第一口时,真是惊喜万分,婴儿食品好吃极了!既营养丰富,又非常健康。婴儿食品是糊状的,所以很容易进食和消

However, the results were evident, and it soon became natural and easy. All my friends laughed at me for eating baby food, but I was not embarrassed to bring out the jars in their company. After all, if a beautiful girl like the flight attendant was not uncomfortable eating baby food in public, why should I care? Anyone making fun of me for doing something that is good for my health is amusing to me anyway. Frankly, I usually make the first joke about eating baby food because it is always fun to laugh about yourself. I never make fun of other people because I believe that is disrespectful and just plain wrong to humiliate anyone else. If you cannot laugh at yourself, you can never truly laugh.

As a side note, if you search the internet for baby food diets, you will find many short-term fad diets that purport to be quick-fix solutions because losing weight is not fun. In reality, those instant-result claims don't last. The long-term answer to being thin for the rest of your life is portion control. And that's what *The Baby Food Diet* is all about.

化。节食八周以后，我的体重减轻了。我是一个崭新的人了！更绝的是，这次体重没有反弹！谢天谢地，**婴儿食品节食法**由此诞生了。

就像所有的节食方法一样，前两周都很难受。然而结果却很明显，并且很快就变得自然和容易了。所有朋友都嘲笑我吃婴儿食品，但我并不为在他们面前拿出婴儿食品罐而感到尴尬。毕竟，如果像空姐那样漂亮的女孩，在公共场合吃婴儿食品都不觉得别扭，我又何必在意呢？任何人因为我做了有益于自身健康的事而取笑我的，对我来说都很好笑。坦白地说，我通常会是第一个对吃婴儿食品开玩笑的人，因为自嘲总是有趣的。我从不取笑别人，因为我认为这样不尊重人，羞辱别人是完全错误的。如果你不能自嘲，你就永远不会真正地欢笑。

附注一点，如果你在网上搜索"婴儿食品节食法"，你会发现很多短期的时尚节食法声称其为快速解决方案，因为减肥很烦人。事实上，那些所谓立竿见影的方法其结果并不能持久。在你的余生中保持身形苗条的长期答案是控制饮食分量，而这正是**婴儿食品节食法**的意义所在。

CHAPTER 2

Three Variations of the Diet

第二章

节食的三个变化阶段

CHAPTER 2
Three Variations of the Diet

There are three variations and phases of *The Baby Food Diet*:

First is the Baby Food Fast which is designed for quick weight loss. While on the Baby Food Fast, you can expect to lose 10 to 15 pounds over two weeks. The Baby Food Fast is not to be followed for more than 14 to 21 continuous days at any one time. (See exact menu on page 82).

Second is the Baby Food Reduction Diet which is the core diet for losing weight steadily. If you are not in a hurry to lose weight, you can skip the Baby Food Fast and start with the Baby Food Reduction Diet. It is simple to follow and should be continued until you reach your target weight no matter how long that takes. On this diet, you will slowly lose your excess weight and easily reach your target goals. The Baby Food Reduction Diet is very healthy. Generally, you can stick to this diet for anywhere between 6 weeks and 25 weeks. (See exact menu on page 82).

Third is the Baby Food Maintenance Diet which is designed to be your forever diet once reaching your goals. With the Baby Food Maintenance Diet, you will never again experience the yo-yo phenomenon common in other diets. (See exact menu on page 82).

Whichever of the three variations you choose, *The Baby Food Diet* is the perfect regimen to keep you lean and

第二章
节食的三个变化阶段

婴儿食品节食法有三种变化阶段：

第一阶段是婴儿食品禁食期，专为快速减肥而设计。在婴儿食品禁食期，你可以预期在两周内减掉 10～15 磅。任何时候坚持婴儿食品禁食期一次不可超过 14～21 天。（具体食谱详见第 83 页）

第二阶段是婴儿食品减食期，这是可以稳定减肥的核心节食时段。如果你不急于减肥，可以跳过婴儿食品禁食期，直接开始婴儿食品减食期。这非常简单易学，而且应该坚持下去，不管需要多长时间，直到达到你的目标体重。在这一阶段，你会慢慢地减掉多余的体重，并且很容易达到自身目标。婴儿食品减食期非常有益健康，一般来说，这一阶段可以坚持 6～25 周。（具体食谱详见第 83 页）

第三阶段是婴儿食品维持期，旨在一旦达到自身目标，它将成为你永远的节食阶段。有了婴儿食品维持期，你将再也不会经历在其他节食方法中常见的"溜溜球"现象。（具体食谱详见第 83 页）

无论你选择哪一阶段，**婴儿食品节食法**都是保持身材

healthy.

As for the actual baby food, there are many varieties. You may not like them all, but most are very tasty. Also, baby food containers are easy to carry during the day and most do not need to be refrigerated. That's another practical reason why *The Baby Food Diet* is much easier to be faithful to than any other diet in the world.

The only drawback of *The Baby Food Diet* is that it takes a few days for your digestive system to adjust. Actually, this is true with all diets. However, the discomfort will soon pass and your body will adapt to your new eating habits, and it will become natural and easy. In fact, it will not be long before you experience a much higher energy level than you were used to in the past. You will also experience so many other amazing benefits, like better sleep, sharper mental focus, and a more positive attitude. Of course, the most obvious difference is your svelte appearance. How much fun it is to look in the mirror and love your new image! And don't be surprised to hear those around you make comments that your persona is laced with sunshine!

The Baby Food Diet does something that all diets preach: establish portion control. Baby food comes pre-measured, so serving size is automatic which means the number of calories consumed is consistent and offset by the

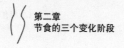

第二章
节食的三个变化阶段

和健康的完美方法。

至于真正的婴儿食品,有很多品种。你可能不会全都爱吃,但大多数都非常美味。此外,婴儿食品的容器在白天也很容易携带,大多数并不需要冷藏。这也是为什么**婴儿食品节食法**比世界上任何其他节食方法更容易遵从的另一个实际原因。

婴儿食品节食法的唯一缺点是你的消化系统需要用几天的时间进行适应。事实上,所有的节食方法都是如此。然而,这种不适很快就会过去,你的身体会适应新的饮食习惯,并且会变得自然和容易。事实上,不久你就会经历比以前更高的能量水平。你还会体验到许多其他令人惊奇的好处,比如更好的睡眠、更专注的注意力和更积极的态度。当然,最明显的区别是你那非常苗条的身材。当你照镜子并喜爱自己的新形象时,该会多有趣!而且在听到周围的人说你整个人都散发着光芒的时候,千万不要惊讶!

婴儿食品节食法也会做所有节食方法都宣扬的事情:制定饮食分量控制方案。婴儿食品是预先测量好的,所以

number of calories burned. Given this crucial constant, *The Baby Food Diet* is surely the world's only perfect diet.

Baby food is rich in vitamins and provides everything that your body needs apart from protein which can be added with supplements, lean red meats, fish, or chicken. Eggs are another option. For example, bring a hard-boiled egg with you as a source of protein. Low-fat milk and low-fat dairy products are another way to add protein. People who are vegetarians can also get protein from sources such as soybeans, soymilk, beans, and lentils. When it comes to protein, change your source on different days to add variety to meals. If you want the fastest results and the most convenient solution, then a high-quality protein powder with a complete amino acid profile is a wise choice. You can also add salads to the diet. For certain social events where you feel the need to assimilate with the people around you, consider a small piece of fish and a green salad.

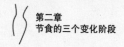

第二章
节食的三个变化阶段

分量是自动的,这意味着摄入的卡路里数量是一致的,并且会被燃烧的卡路里数量所抵消。考虑到这一关键因素,**婴儿食品节食法**无疑是世界上唯一完美的节食方法。

婴儿食品富含维生素,并且除了蛋白质以外,可以提供身体所需的一切,而蛋白质可以通过营养补充剂、瘦肉、鱼或鸡肉补充,鸡蛋是另一种选择。例如,随身带一个煮熟的鸡蛋作为蛋白质的来源,低脂牛奶和低脂乳制品也是另一种添加蛋白质的方式,素食者还可以从大豆、豆浆、豆类和扁豆等来源处获得蛋白质。说到蛋白质,在不同的日子可以改变不同来源以增加食物的多样性。如果你想要最快的结果和最方便的解决方案,那么高品质的蛋白粉与完全氨基酸搭配是一个明智的选择;你也可以在饮食中添加沙拉;在某些社交场合,当你觉得需要融入周围之人时,可以考虑吃一小块鱼肉和一份蔬菜沙拉。

CHAPTER 3

Knowing Your Weight

第三章

自知体重

CHAPTER 3
Knowing Your Weight

We all know that being thin is healthier than being fat. *The Baby Food Diet* will not only keep you lean and your energy will explode making you more productive throughout the day. We have all heard how drinking a lot of alcohol and smoking cigarettes will shorten your life. I am not a doctor, but being overweight seems as obvious and detrimental to your health.

As mentioned, *The Baby Food Diet* is a forever regimen. Once you reach your target weight, you must keep on eating baby food, but only one meal per day. The odd thing is that I never get tired of ingesting baby food. Think about that statement. When I started the diet the first time, I cringed at the thought of swallowing such food. Now, I enjoy it daily. Another benefit of the diet reflects on your savings account. Even high-quality baby food is cheaper than most other food choices, so you will also save money!

Even if you develop an insatiable appetite for baby food, you cannot consume as much as you desire. It must be limited to two or three portions each day if you wish to lose weight and keep it off. With that in mind, there might be a day where you have baby food for breakfast, a protein-shake for lunch, fish and another baby food for dinner. If you want faster weight loss, then you can enjoy baby food for breakfast, a protein-shake for lunch, and another baby food for dinner.

第三章
自知体重

我们都知道瘦比胖更健康。**婴儿食品节食法**不仅会让你保持身材,旺盛的精力会让你感觉在一整天中更有效率。我们都听说过大量饮酒和吸烟会缩短寿命。我不是医生,但是超重对于健康显然也是一样有害的。

如前所述,**婴儿食品节食法**是一个永久方案。一旦达到目标体重,你仍然必须继续进食婴儿食品,但每天只有一餐。奇怪的是,我从来没有厌倦过吃婴儿食品。想想当初我说的,当我第一次开始节食时,一想到要吞下这样的食物,我就畏缩不前。而现在,我每天都很享受。节食的另一个好处体现在你的储蓄账户上。即使是高品质的婴儿食品也比大多数其他食品便宜,所以你也会因此省钱!

即使你对婴儿食品产生出贪得无厌的胃口,你也不能想吃多少就吃多少。如果你想减肥并保持体重,每天必须控制在2~3份。要记住这一点,在一天之中也许你可以早餐吃婴儿食品,午餐吃蛋白奶昔,晚餐吃鱼和另一份婴儿食品。但如果你想更快地减肥,那么你可以享用婴儿食品当早餐,午餐是蛋白奶昔,另一份婴儿食品当晚餐。

For days when you are extra hungry, a third baby food is acceptable for an afternoon or evening snack. It is said that never to eat after 7 pm if you want to lose weight. The truth is if you are hungry late at night, baby food is your answer due to its pureed form and easy digestion.

As is true on all diets, drinking water is super important. Your body is mostly made up of H_2O. When you are on *The Baby Food Diet*, you should drink at least eight glasses of water per day as recommended by health experts. As a matter of fact, on *The Baby Food Diet* I never seem to get the urge to gorge. However, if you also drink lots of zero calorie fluids in addition to baby food, you will never be hungry. That's right, I said you will *never* be hungry on *The Baby Food Diet*. Some suitable beverages include any kind of black coffee or tea without adding milk or sugar. Plain soda water, like Perrier, is also excellent at making you feel full when dieting. I often drink an Americano.

When beginning *The Baby Food Diet*, I suggest weighing yourself first. Tracking your weight is important, so I recommend calculating your body mass only once per month because weight does not come off in a linear fashion. There will be times when you are doing everything right and still gain a pound or two. This type of unexpected weight increase is just

第三章
自知体重

当你特别饿的时候,第三份婴儿食品可以作为下午或晚上的零食。据说,如果你想减肥,晚上7点以后绝对不要吃东西。事实上,如果你在深夜很饿,婴儿食品是你的最佳选择。这主要在于其为糊状且易于消化。

就像所有节食方法一样,喝水是非常重要的,人的身体主要由 H_2O 组成。在开始**婴儿食品节食法**时,按健康专家建议,你应该至少每天喝8杯水。事实上,在**婴儿食品节食法**期间,我似乎从未有过狼吞虎咽的冲动。然而,如果除婴儿食品外,你还喝很多零卡路里的饮品,就根本不会感觉到饿。没错,我说了,**在婴儿食品节食法**期间你根本不会觉得饿。一些合适的饮品包括任何不加牛奶或糖的黑咖啡或茶,普通苏打水,比如毕雷矿泉水(Perrier),也能有很好的效果,让你在节食时有饱腹感。我则是经常喝美式咖啡。

我建议,在开始**婴儿食品节食法**之前,先量一下自身体重。关注自身体重很重要,所以我建议每月只检查一次自己的体重,因为体重不会呈线性方式下降。有时候即使做的每件事都正确,但你还是会增重1~2磅。这种意想不到的体重增加只是减肥过程中的一部分,

part of the process of shedding excess, nonetheless, it can be depressing. We all know that salt causes water retention which can add significant weight. Many other factors, especially stress, can increase heft fluctuations. With the above variables, weighing only once per month is suggested. Rest assured on this diet at the end of every month, you will have lost some weight. How much fun is it stepping on the scale and seeing that you have lost some weight? It makes me feel like I accomplished something when the scale shows positive results.

Every diet book agrees that individuals should not be in a hurry to become thin because each person has a different metabolism. I will echo the same caution with *The Baby Food Diet*, although you will drop weight faster than you can imagine, it still takes time. You won't even need a scale to see your weight loss because your clothes will be falling off!

I have a proven antidote for maintaining your ideal poundage after successfully losing weight. Once your target weight is achieved, you must continue weigh yourself, but only once every thirty days. At the end of each month, if you have gained 5 pounds or more, it is time to go back to the Baby Food Reduction Diet. On the other hand, if your weight is up 4 pounds after a month, don't worry about it. In such a case, wait another month and if you have gained 5 or more pounds over your goal

第三章
自知体重

但尽管如此,可能还是会令人沮丧。我们都知道盐会导致水潴留,从而增加体重。其他许多因素,特别是压力,可以增大体重波动。由于上述变量,建议每月只称重一次。请放心,按照这种节食方法,在每个月的月底,你的体重都会减轻一些。当你踏上体重秤,发现自己变轻了,该是多么有趣啊!当体重秤显示出积极的结果时,我会感觉自己很成功。

每一本节食方面的书籍都认为,人体不应该急于变瘦,因为每个人的新陈代谢都不一样。在**婴儿食品节食法**中,我也会重复同样的警告,尽管减肥的速度比你想象得要快,但仍然需要时间。你甚至不需要体重秤就能感受到自己的体重减轻了,因为你的衣服会变得宽松起来!

我有一种经过验证证实行之有效的方法,可以帮助你在成功减肥后保持理想的体重。一旦目标体重达到了,仍要继续称自己的体重,但是每30天只称重1次。在每个月月底,如果你的体重增加了5磅或更多,是时候回到婴儿食品减食期了。另一方面,如果你的体重在一个月后增加了4磅,不要担心,在这种情况下,再等一个月,如果你的体重比目标体重增加了

weight, then more heavy-handed changes are in order.

When this happens, simply restart the Baby Food Fast. At the end of the next month, weigh yourself again. At that point, if you have returned to the goal weight, you can go back to the Baby Food Maintenance Diet. If you are still above your target weight, return to another month on the Baby Food Reduction Diet. If you keep having to go back and forth between the Baby Food Fast or Baby Food Reduction Diet, then you need to take a close look at your maintenance diet and portion intake.

This brings us to the age-old practice of keeping a food journal. When you are on the initial Baby Food Reduction Diet, there is no need for such a record. You are doing the same basic thing every day with some slight modifications. However, on the Baby Food Maintenance Diet, when you are eating two normal meals and substituting the third meal with baby food, keeping a journal is a good idea. If you continually experience weight gain beyond 5 pounds in a month while on the Baby Food Maintenance Diet, the journal is definitely a must. The journal enables you to clearly see when and where you are going wrong on your maintenance food.

Is a journal 100% necessary? The answer is no. You can control your weight by simply planning a Baby Food Fast or Baby Food Reduction Diet every now and then to trim down. However, if you are experiencing this

第三章
自知体重

5磅或更多，那么就该进行更严格的调整。

当这种情况发生时，只需快速重启婴儿食品禁食期，下个月月底，再称一次体重。到那时，如果你已经回到了目标体重，你可以重新回到婴儿食品维持期；如果你的体重仍然超过目标，再用一个月的时间回到婴儿食品减食期；如果你不断在婴儿食品禁食期和婴儿食品减食期之间来回切换，那么你需要仔细检查维持期进食的食物及其摄入量。

这就让我们重新回到写食品日志这种老式做法上来了。在开始婴儿食品减食期的时候，不需要这种记录。每天都是按照基础的方案进食，只是做一些小小的修改而已。然而，在婴儿食品维持期，每天吃两顿正常的饭，并用婴儿食品代替第三顿饭时，记日志是个好主意。在婴儿食品维持期期间，如果你不断经历体重在一个月内增长超过5磅，那么记日志绝对是必需的。这本日志能让你清楚地看到何时、何地所进食的食物出了问题。

日志是百分之百必要的吗？答案是否定的。你可以简单地通过婴儿食品禁食期或婴儿食品减食期来控制自身

fluctuation problem, I believe committing to a journal can be a beneficial solution for monitoring your maintenance problems. When consulting the journal, you may find it necessary to only change one or two small eating habits to bring your weight under control. No matter how you choose to maintain your desired physique, *The Baby Food Diet* keeps you thin and looking fabulous.

You may be asking, "What expectations should I have with *The Baby Food Diet* regarding how much weight I can lose, and how quickly I can lose it?" The answer is this diet is so easy to stay on that you will crave for nothing. Weight loss is not the same for every person as there are multiple factors that enter the equation. How fast is your metabolism? How much exercise do you get? How much stress do you encounter in your daily life? How much sleep do you get? I lost about 10 pounds in two weeks when I first went on the Baby Food Fast. Some people lose more, and some lose as little as 5 pounds during the same period. Think about it, even if a person only loses 5 pounds each month, it still adds up to 60 pounds each year. If you add exercise to the equation, you will typically lose a minimum of 10 pounds in a month. That is 120 pounds a year.

Another factor is how overweight are you currently? The heavier you are, the quicker weight will fall off. As you

第三章
自知体重

体重，不时地进行调整。但是，如果你正在经历这种波动问题，我相信写日志对于监控日常维护问题是非常有益的解决方案。当你查阅日志时，会发现只要改变其中一两个小的饮食习惯就能控制自身体重。无论你选择怎样的方式来保持理想的体形，**婴儿食品节食法**都会让你保持身材且赏心悦目。

你可能会问："关于能够减掉多少体重，我应该对**婴儿食品节食法**抱有怎样的期望，以及能以多快的速度减重？"答案是，这种节食方法很容易坚持下去，你不需要渴望什么。减肥对每个人来说都不一样，因为有很多因素在起作用。新陈代谢有多快？做了多少运动？日常生活中遇到多少压力？每天有多长时间的睡眠？当我第一次开始婴儿食品禁食期的时候，在两周时间内瘦了大约10磅。有些人减掉的更多，而有些人同期只减掉了5磅。想想看，即使一个人每个月只减5磅，每年加起来还是能有60磅。如果在此基础上再加上运动，通常一个月内至少会减掉10磅，一年就是120磅。

另一个因素是你现在体重有多重？体重越重，减得越快。当你越接近目标体重时，减得越慢。这会有什么问

get closer to your target weight, the pounds come off more slowly. Does it matter? I say it does not matter at all. *The Baby Food Diet* is a continual lifestyle strategy, so you will never feel like you're missing out on anything. Whether fast or slow, you will lose the desired weight and eventually reach your target size. This is the only diet that can boast such a truth.

第三章
自知体重

题吗？我说根本没有任何问题。**婴儿食品节食法**是一种持续的生活方式策略，所以你根本不会觉得少了什么。无论是快还是慢，你都会减掉理想的重量，最终达到目标体形。这是一种真正有效的节食方法。

CHAPTER 4

Exercise and Health

第四章

运动与健康

CHAPTER 4
Exercise and Health

I believe exercise is important, but is not required as a part of *The Baby Food Diet*. Obviously, working out improves everyone's overall health. But getting fit does not require tedious hours in the gym every day. Sure, if you have time and enjoy it, exercising regularly is key to maintaining good health.

Breaking a sweat helps you experience positive physical and mental health changes in your lifestyle and can be done outside the gym. For example, take the stairs instead of the elevator. When watching television, ride a stationary bike. Do sit-ups in your bed before falling asleep. Go for a walk every day. Join social activities like dancing or yoga. Every extra little thing you do can make a difference in the speed you lose weight. Regardless, you will reach your target weight by following *The Baby Food Diet* whether you exercise or not. I love to work out and stay active, but I realize that many people do not enjoy exercise. There are people who live to be over 100 years old and have never engaged in regular physical activity.

We all love to be complimented. It feels good to be noticed in a positive way, and that is exactly what will happen with *The Baby Food Diet*. Your friends and family will see a new you. In little to no time, you will take on a more positive approach to everything you do and exude strength and confidence that affect every

第四章
运动与健康

我认为运动很重要,但这不是**婴儿食品节食法**所强求的部分。显然,运动可以改善每个人的整体健康。但是,瘦身并不需要每天在健身房耗上几个小时。当然,如果你有时间并且乐享其中,经常运动同样是保持健康的关键。

流点汗可以有助于你在生活方式上体验到积极的身心健康变化,而且不用去健身房也一样可以有效果。例如,走楼梯而不是乘电梯;看电视时,脚踏动感单车;睡觉前在床上做仰卧起坐;每天散步;参加社交活动,如跳舞或做瑜伽。你所做的每一件额外的小事都会对减肥的速度产生影响。无论是否运动,只要遵循**婴儿食品节食法**,你就能达到自身目标体重。我喜欢健身和保持活力,但我意识到很多人并不喜欢运动,有些人活到100岁以上,却从未进行过任何有规律的体育活动。

我们都喜欢被人称赞。能以积极的方式被别人注意到,这种感觉很好,这正是**婴儿食品节食法**将会带来的好处。你的朋友和家人会看到一个全新的你。很快,你就会对所做的每件事采取更加积极的态度,并散发出能影响自身生活方方面面的力量和自信。你看待事物

aspect of your life. The way you look at things will be different, more optimistic. Of course, your health will improve greatly. In most cases blood pressure will drop, and the fat in your liver will be reduced as a direct result of *The Baby Food Diet*. The benefits are so far-reaching and life-changing that it's influence on your life can never be fully measured.

Another advantage of *The Baby Food Diet*, at least for me, is that it has cured the annoying acid reflux problem I used to have. Before partaking in *The Baby Food Diet*, I was constantly taking medication to control my sour stomach. Now, I am free from popping Rolaids, Tums, Prilosec, etc. My stomach is always balanced, and I never think about heart burn anymore.

Here is a typical day on the road to weight loss with *The Baby Food Diet*. When I was losing the weight, I would wake up and have a protein shake. At lunch, I would have my first baby food. In the afternoon, but not every day, I would have a second baby food snack if I felt hungry. Dinner is where I would change it up. I am a businessman, so I must often dine out to discuss transactions. This means going to a variety of restaurants. In this situation, I have the second baby food of the day. In addition, I might order a small salad with vinegar dressing and some form of lean protein. Normally, I choose a small

第四章
运动与健康

的方式将会不同,会更加乐观。当然,你的健康会大大改善。大多数情况下,血压会下降,肝脏中的脂肪会减少,这是**婴儿食品节食法**的直接结果;其好处是如此深远,生活也因此而改变,它对生活的影响永远无法充分估量。

婴儿食品节食法的另一个好处,至少对我来说,是它治愈了我曾经讨厌的胃酸倒流问题。在开始**婴儿食品节食法**之前,我一直靠服用药物来控制胃酸。而现在,我不需要再服用碳酸二羟铝钠(Rolaids)、胃酸片(Tums)、奥美拉唑(Prilosec)等药物了,我的胃现在一直很平衡,再也没有烧心的感觉了。

这是尝试**婴儿食品节食法**减肥过程中典型的一天。在瘦身过程中,我会早上醒来喝一杯蛋白奶昔。午餐时,我会吃第一份婴儿食品。在下午,但不是每天如此,只在觉得饿了时,吃第二份婴儿食品当零食。晚餐我会有所改变。我是商人,所以必须经常外出吃饭谈生意,这意味着要去各种各样的餐馆。在这种情况下,我会吃一天中的第二份婴儿食品。此外,我可能会点一小份浇上醋汁的沙拉,外加一些含蛋白质的瘦肉。通常,我会选择一小份鸡肉或鱼肉。

portion of chicken or fish. The diet is that easy and that simple. Controlled portions with some flexibility built in. If I was not having a business dinner, then I would stay home and have a baby food and possibly a salad or fish.

Not to fear, there will be some delicious meals which complement the baby food and protein supplements. Except during the Baby Food Fast, you will also be consuming various normal foods. The rule of thumb is to count calories and implement portion control for these richer meals. Following this advice is imperative to success on the diet.

However, our goal is to maximize weight management success, not to merely help you become skinny. We also want you to be healthy and enjoy a balanced and flavorful diet. To achieve this goal, we have recruited highly trained nutritionists who have been certified as "Baby Food Diet" experts. These professionals have designed numerous recipes that taste sumptuous while helping maintain a healthy, slender body. (They can be accessed on our app and website by chat, tele/video conference, or in-person that will assist you with the diet and offer recommendations. You can ask basic questions regarding meals, ingredients, nutritional pros and cons, anything of importance to you. Of course, no medical advice will be given, for that you must ask your personal physician.)

第四章
运动与健康

饮食就是如此简单方便,控制饮食分量并加入一些灵活性。如果我不去吃商务晚餐,则会待在家里吃婴儿食品,可能还会吃沙拉或鱼。

不用担心,会有一些美味的食物辅助婴儿食品和蛋白质营养补充剂。除了婴儿食品禁食期,还可以吃到各种正常的食物。经验法则就是计算卡路里并控制这些丰盛食物的分量。遵循这一建议是成功节食的必要条件。

然而,我们的目标是将体重管理的成功最大化,而不仅仅是帮助你变瘦。我们还希望你能够健康,享受均衡和美味的饮食。为了达到这一目标,我们聘请了训练有素的营养学家,他们被认证为"婴儿食品节食法"专家。这些专业人士设计了许多美味的食谱,可以帮助我们保持健康、苗条。(你可以在我们的应用程序和网站上通过聊天、电话/视频会议或亲自访问等方式与他们交流,来帮助你节食并提供建议。可以问一些基本问题,关于食物、配料、营养方面的利弊,任何对你来说重要的事情都可以。当然,他们并不提供医疗建议,因为这方面你必须询问自己的私人医生。)

CHAPTER 4
Exercise and Health

Our nutritionists are independent, third party professionals not paid or employed by us. In fact, if you are a nutritionist who wishes to be included on this list, please forward us your relevant information. After we review your credentials and provide our guidelines, you may qualify to be included in this prestigious group.

You should realize that protein powder is normally flavored with artificial sweeteners, so your desire for confectioneries will most likely be satisfied. However, if you have an occasional sweet tooth, then put two Hershey Kisses in your bag and enjoy them when needed. They are very low in calories, but most importantly, this is an example of portion control at its finest. There are other small snacks like this, but the point is to plan ahead when considering serving sizes. We all know reaching into a big bag of potato chips is entering the danger zone. One potato chip company used to have an advertisement saying: "I dare you to eat just one." The ad is accurate as it is hard to eat just one of anything that is so tasty.

The solution to potentially overeating high calorie foods is simple. For example, I take five potato chips and put them in a separate bag. That separation prevents me from gorging. It retrains my brain to enjoy small portions. With the big bags, even the most disciplined person will overindulge. You've got to set up consumption parameters

第四章
运动与健康

我们的营养学家是独立的第三方专业人士,不是我们出钱雇用的。事实上,如果你是一名营养学家,想要被列入这份名单,请把相关信息转发给我们。在我们审核完你的资质并提供我们的指导方针后,就能有资格被纳入这个享有盛誉的团体。

你应该能意识到,蛋白粉通常是用人造增甜剂调味的,所以你对糖果的渴望很可能会由此得到满足。不过,如果你偶尔嘴馋想吃甜食,那就在包里放两块好时巧克力(Hershey),在需要的时候享用,其卡路里含量很低,而最重要的是,这是一个最好的控制饮食分量的例子。还有其他类似的小零食,但重点是在考虑食用分量时要提前规划。我们都知道,把手伸进一大包薯片里就等同于进入了危险区。一家薯片公司曾经做过这样一则广告:"你敢吃一片吗?"这则广告是准确的,因为面对如此美味的食物,仅仅只吃一片实在太难了。

解决可能过量摄入高热量食物的方法很简单。例如,我拿出 5 片薯片,把每片放在一个单独的袋子里。这种分别开来的方法使我不会一次进食过多,能够重新训练我的大脑去享受小份。有大袋食物在手,即使是最自律的

for success.

This strategy of separating food into small portions allows you to still enjoy a few guilty pleasures. As far as I'm concerned, it is okay to cheat on a diet. In fact, small slip-ups can be fun. Sometimes even a huge slip-up is totally acceptable. The secret is to get back on track with the diet to ensure that occasional eating blunders have minimal impact on your goal of being lean and fit.

Another good stimulus to stay on track is to hire an artist to draw a picture of how you would look at your new, thinner target weight. Look at the rendition every day. Copy the image and put it in your bathroom, your kitchen, your car, your office and everywhere you go. Your subconscious mind will make the picture a reality. Remember, what you put in your mind, what you think, becomes reality.

One of the reasons a diet is hard to begin is fear. The fear of failure and embarrassment are enemies to progress. Fear is a funny thing. If not confronted, it will fester and destroy you. Besides, why should you ever be embarrassed for trying to improve yourself? Even if you fall off the diet, why is that embarrassing? Really, isn't the only real embarrassment *not* trying something? People focus too much on failing, not on success. It should be the opposite. If you at least try, you have already succeeded

第四章
运动与健康

人也会过度放纵。你必须为成功设置进食界限。

这种把食物分成小份的策略，让你仍然可以享受一些内疚的快乐。在我看来，在节食中"作弊"是可以的。事实上，小小的过失也会很有趣。有时候，即使是一个巨大的过失也是完全可以接受的。秘诀就是再回到正常节食的轨道上来，以确保偶尔的饮食错误对自己保持身材和健康的目标影响得最小。

另一个能激励你坚持走在正确轨道上的好方法，是聘请一位艺术家来为你画一幅画，描绘出你达到目标体重时，那个崭新的、更纤瘦的自己。每天看看画中演绎出的自己，复制这张画，把它放在浴室、厨房、车内、办公室以及任何你会去的地方。你的潜意识会使这幅画成为现实。要记住，你的所念所想，都会变成现实。

恐惧是节食难以开始的原因之一。害怕失败和感到难堪是前进的敌人。恐惧其实挺有意思，如果你不抗争，它就会恶化并摧毁你。此外，你为什么要为自己的进步而感到尴尬呢？即使你节食失败了，为什么会感到尴尬呢？真的，不敢尝试某些东西不是唯一应该感到

just by making the attempt. To me, if I try something, there is no such thing as failure.

People often use the reason of becoming healthier to start a diet. Sure, as a result of dieting you will most certainly be healthier. However, it is unlikely that this motive alone will be enough unless you had a health scare. You need real motivation that clicks within you, not whimsical, fleeting inspiration.

So how do you find an incentive to lose weight that will become your passion and obsession? For many people, it is the desire to look their very best to enhance opportunities to find the love of their life. For others, it is an upcoming event where they want to impress attendees with their stunning appearance. For yet others, it is a health scare.

It is never easy finding a strong motivation to coerce you into seriously losing weight, but it is important. I recommend sitting down and writing all the ways your life will be improved once you are thin. Maybe one of them will click. If this does not work, then make it a game and play to win by keeping score. Dieting with another person to make it competitive is something I have seen work. Maybe rewarding yourself with a gift will work as an incentive, like a new wardrobe or possibly that piece of jewelry you always wanted but could never justify

第四章
运动与健康

尴尬的事吗？人们过于关注失败，而不是成功。其实恰恰应该相反。如果你至少尝试一下，就已经算是一种成功。对我来说，如果我尝试一件事，就没有失败这回事。

人们经常以"能够更加健康"的理由来开始节食。的确，节食的结果一定会使你更健康。然而，除非你曾有过健康危机，否则单凭这个动机是不够的。你需要的是发自内心的真正动力，而不是异想天开、心血来潮的想法。

那么，如何才能找到一个减肥的动力，使它成为你的激情并痴迷其中呢？对很多人来说，是渴望自己的形象看起来更好，从而增加找到一生挚爱的机会；而对另一些人来说，是在即将到来的某一活动中，想通过自己惊人的外表给与会者留下深刻印象；对还有些人来说，是因为健康危机。

能找到一个强大的动机来强迫自己认真减肥，从来都不是件容易的事，但这很重要。我建议你坐下来，写下所有在你瘦身以后生活将会改善的方面。也许其中有一个会击中要害。如果这不起作用，那就把它变成一场比赛，通过记分来定输赢。和另一个人一起节食，

purchasing. Whatever it takes to stay on a diet, it will get easier, but that initial motivation is vital.

Willpower is one of the keys to a successful diet. Being able to say no to temptation is a must on your road to success. Developing and maintaining routines and rituals are powerful tools for strengthening your willpower. It seems that following a routine, such as doing the same activities, at the same time each day increases your odds of success. Developing positive habits is a common trait found in successful people. Maintaining a steady set of habits is very helpful for maintaining willpower. Once you have a routine of strong habits built into your daily activities, there won't be as many decisions that can derail your willpower. For example, going to the gym at the same time each day is a good habit to follow. This routine and habit will increase the odds that you go on a regular basis. Below are more thoughts on willpower.

Being on a diet is like running a marathon, as both require stamina. In a long race, taking one step at a time will bring success. It is a must to keep going and not stop. Marathon runners have a fear of hitting the wall and not finishing. This fear drives the runner to keep going. We call this a healthy fear. Training for a marathon will build willpower as a result of the daily training. Then completing the race, despite your mind telling you to stop, empowers the runner and spills this success over to other

第四章
运动与健康

进行比赛,是我见过的行之有效的方法。也许奖励自己一件礼物也会起到激励作用,比如买一个新衣柜,或者买一件你一直想要但永远找不出购买理由的珠宝。无论哪种,只要能让你坚持节食,事情就会变得简单些,但最初的动力是至关重要的。

意志力是成功节食的关键之一。对诱惑说不,是通往成功的必经之路。培养和保持日常习惯是增强意志力的有力工具。按照习惯,每天在同一时间做同一件事似乎会增加你成功的概率。培养积极的习惯是成功人士的一个共同特征。保持一套稳定的习惯对保持意志力很有帮助。一旦你在日常活动中养成了一些坚定的习惯,就不会有那么多的决定能让你的意志力偏离轨道。例如,每天在同一时间去健身房是一个好习惯。这种惯例和习惯会增加你定期去的可能性。下面是关于意志力的更多想法。

节食就像跑马拉松,两者都需要耐力。在长跑中,一步一个脚印会带来成功。一定要坚持往前跑,绝不能停。马拉松运动员最害怕撞墙(注:撞墙为马拉松术语,意为突然跑不动了)而不能完成比赛。这种恐惧驱使跑者继续前进。我们称之为健康的恐惧。在马拉松

areas of the runner's life. Sticking with the baby food diet will produce a similar experience.

A fantastic way to build will-power is to take cold showers. A short cold shower will strengthen your willpower. Small amounts of suffering on your journey to becoming thin will increase your willpower. Going to the gym has taught me, no pain and no gain. After a cold shower you will feel empowered, refreshed and confident.

When exhaustion has brought you to the breaking point, stop everything. Take a break and relax your mind. Go for a walk in the park, climb to the top of a mountain and scream at the top of your lungs. Undertake activities that activate your brain. Maybe go to a movie, meditate, talk to a stranger, the list is endless. The point being, taking time away from responsibility can indeed be a magic elixir that will allow you to push forward. When your computer or phone freezes, what is the solution? The answer is to reboot, this also works in real life.

Building willpower and productivity requires taking control of your actions and thoughts. It is a must to remain committed to your short and long-term goals. Never taken action without first slowing down and thinking. When you become frustrated, keep calm and get a clear picture of what needs to be resolved. With this approach you will avoid making emotional decisions.

第四章
运动与健康

训练中,日常训练的结果就是培养意志力,然后完成比赛,尽管大脑告诉你停止,却依然赋予长跑者以力量,并把这种成功复制到长跑者生活的其他领域。坚持**婴儿食品节食法**也会产生类似的体验。

培养意志力的一个绝妙方法就是洗冷水澡。洗个短暂的冷水澡会增强你的意志力。在变瘦的过程中,一点点的痛苦会增强你的意志力。去健身房让我明白,一分耕耘一分收获。洗完冷水澡后,你会感到精力充沛、神清气爽、充满自信。

当你疲惫得到了崩溃的边缘时,停止一切,休息一下,放松大脑。去公园散散步,爬到山顶,使出所有肺活量大声叫喊。做一些能激活大脑的活动,也许去看场电影,练习冥想,和陌生人聊天,这样的活动不胜枚举。关键是,花些时间远离责任确实可以成为一剂神奇的灵丹妙药,能让你继续前行。当你的电脑或手机死机时,怎么解决?答案是重启,这在现实生活中也同样是可行的。

要培养意志力和创造力需要控制你的行为和思想。坚持短期和长期目标是必须的。如果不先放慢脚步来思

Also, why not identify and remove distractions, so you can stay committed to your plan. For example, do not keep unhealthy foods in your home.

You can train your mind to use willpower as a catalyst to achieve the bigger goals that you want to achieve. Start by taking one step past your limits, then two and then three. This will allow your brain to reprogram itself. Before long you will experience huge leaps beyond what you thought was possible.

A trick I use when reaching a point of weakness is to just say "no" out loud. Whether you scream it or whisper it does not matter. For some reason, the act of voicing the "no" is a powerful tool in maintaining your willpower. For some strange reason this works. Anything you can do to get past the point of weakness will get you closer to your goal of being thin and healthier.

第四章
运动与健康

考,就不要采取行动。当你感到沮丧时,保持冷静,要明确什么问题需要解决。通过这种方法,你将避免做出情绪化的决定。此外,为什么不找出并排除干扰,这样就能坚持自身计划。例如,不要把不健康的食物放在家里。

你可以训练自己的大脑,把意志力作为催化剂来实现理想中更远大的目标。从超越自身极限一步开始,然后两步,然后三步,这将使你的大脑重新编程,不久就会经历超出你认为可能的巨大飞跃。

当我遇到自身意志力薄弱时,会使用一个小技巧,那就是大声说"不"。无论你是喊出还是低语,都无关紧要。出于某种原因,说"不"是保持意志力的有力工具。出于某种奇怪的原因,这招真的管用。不管用什么办法,只要能克服自身弱点,就能让你更接近瘦身和健康的目标。

CHAPTER 5

Mindset and Other Interesting Thoughts

第五章

心态和其他
有趣的想法

CHAPTER 5
Mindset and Other Interesting Thoughts

Let's talk a little more about mindset. For me it has always been difficult starting a diet. I will often talk for months about it before actually beginning. Then, once I start, it is a major challenge to not stray for the first two weeks which are the most critical and difficult. I have found that if I can stick to a diet for two solid weeks, it becomes second nature.

This is also true of *The Baby Food Diet*. For the first couple of weeks you, too, will find yourself challenged. You will start thinking about all the favorite foods you are missing. As a result, I developed an interesting mental exercise to overcome this problem. When I find myself craving for a particular treat like a donut, I go ahead and buy it. However, I will not buy just one donut, but a dozen! Then I give them all away to friends and coworkers at my office. I get satisfaction from watching everyone eating the donuts and my urge disappears. For some reason, I no longer want a donut or whatever it was for which I was craving.

It is interesting how the mind operates. I am not saying this will work for you, but it has always helped me curb those insatiable cravings. I am not a psychologist, so I am not sure why or how this works, but it does. I suggest trying this simple ploy next time you get an overwhelming desire to munch something that will blow your diet. Maybe it will be an answer for you, or

第五章
心态和其他有趣的想法

让我们再多谈一谈心态。对我来说,一直都很难开始节食。在真正开始之前,我经常会空谈好几个月。然后,一旦开始,在前两周是否能够不离不弃地坚持会是一个主要的挑战,这是最关键和最困难的。我发现,如果能实实在在地坚持节食两周整,它就会成为我的第二天性。

婴儿食品节食法也是如此。在最初的几周,你也会发现自己面临挑战,你会开始想念所有错过的美食。因此,我开发了一个有趣的心理练习来克服这个问题。当我发现自己特别想吃甜甜圈之类的东西时,我就会去买。然而,我不会只买一个甜甜圈,而是一打!然后我把它们都送给办公室里的朋友和同事。看着他们每个人都吃甜甜圈,我从中获得了满足,自己想吃的欲望就消失了,由此我不再想吃甜甜圈或任何渴望吃到的东西了。

大脑的运作方式非常有意思。我并不是说这会对你有用,但它总能帮我抑制那些贪得无厌的欲望。我不是心理学家,所以不知道为什么会起作用,也不知道怎么起的作用,但它确实起了作用。我建议,下次当你有强烈的吃东西的欲望,并会毁掉节食计划时,可以试

possibly you will eat the whole box of donuts yourself. I hope that's not the case.

Another deterrent to remember is that there is no room for peer pressure when dieting. The same person who is teasing you about being overweight will later try to convince you to cheat. Is it because misery loves company? Is it because some people are just plain mean? Is it because you might soon look better than them and jealousy has entered the picture? The fact is none of this matters and don't let it get in the way of your goals.

As children, we all had someone dare us to do something stupid. Most kids will fall for a prank at least once and then, hopefully, learn their lesson and not be influenced again by peer pressure. Most successful people are confident in their abilities and the path to attaining their goals and are immune to peer pressure. *The Baby Food Diet* will teach you to ignore goading and negativity from all sources because you will quickly see the positive results. Sure, if you want to be spontaneous and occasionally go off the diet, it won't hurt. However, the next day jump back on the diet like that never happened. Make it *your* decision, not a choice based on what others are saying.

In today's world, poor air quality, polluted water, and other harmful environmental necessities cause our bodies

第五章
心态和其他有趣的想法

试这个简单的方法。也许这正是适合你的解决方案,又或者也有可能你自己吃掉整盒甜甜圈,我真心希望不是这样。

另一个需要记住的障碍是,节食时不要理会来自同伴的压力。那个取笑你超重的人,之后会试图说服你在节食时作弊。是因为同病相怜吗?还是因为有些人天生就刻薄?是不是因为你可能很快就会看起来比他们好,因此嫉妒心也加入其中了?事实上,这些都不重要,重要的是不要让它妨碍你的目标。

作为孩子,我们都经历过被他人挑唆做傻事的时候。大多数孩子都至少被恶搞过一次,然后,但愿他们能吸取教训,不再受同伴压力的影响。大多数成功人士都对自身能力和实现目标的道路充满信心,并不会受到来自同伴压力的影响。**婴儿食品节食法**将教会你忽略来自各个方面的挑唆和负面影响,因为你很快会看到积极的结果。当然,如果你在无意中或偶然间中断了节食,也不会有什么影响。无论如何,第二天,再跳回正常节食轨道,就好像之前的事从未发生过。要自己做出决定,而不是选择听从他人之词。

to become full of toxins. Animals are on steroids and vegetables are sprayed with questionable pesticides which cannot be good for your health. *The Baby Food Diet* will help remove these toxins from your digestive system in a natural way.

Baby food is the purest food you can eat. I believe *The Baby Food Diet* helps prevent a myriad of health problems that extend beyond fat related illnesses. It seems logical that if the food you consume is free of contaminants, you will enjoy better health and are less likely to get sick. And on the rare cases when you do become ill, your recovery time will be faster as a result of *The Baby Food Diet*. Again, I have no scientific evidence to support my claims, but to me it just makes sense. Following *The Baby Food Diet* will allow you to improve your overall well-being.

When we look back at historical pictures of people, we see individuals who were considered beautiful, but they may not necessarily appear attractive by contemporary standards. This is due to changing fashions, hair styles, and other trends as time passes. There was a time in history when being fat was considered the most attractive aspect of a person. Even today, in many countries being portly is associated with being rich. However, in most places, being thin means success, and that will likely change as time marches forward. What will not change is that being thin is superior

第五章
心态和其他有趣的想法

当今世界,恶劣的空气、污染的水和其他有害的环境必需品导致我们的身体充满了各种毒素。给动物使用类固醇,给蔬菜喷洒有问题的农药,这些都对健康没有好处。**婴儿食品节食法**将帮助你以自然的方式清除消化系统中的这些毒素。

婴儿食品是你能吃到的最纯净的食品。我相信,**婴儿食品节食法**有助于预防大量的健康问题,而不仅仅限于与肥胖有关的疾病。如果你吃的食物不含污染物,你就会更健康,也更不容易生病,这似乎是合乎逻辑的。而在极少数情况下,当你真的生病了,由于**婴儿食品节食法**的缘故,你康复的时间会更快。同样,我没有科学证据来支持这一观点,但对我来说,这很在理。遵循**婴儿食品节食法**会使你提高整体健康。

当我们回看老照片上的人物时,那些曾经被认为很漂亮的人士,如果以当下的标准来看,却不一定有吸引力。这是因为随着时间的推移,时尚、发型和其他潮流等都在发生变化。历史上曾经有一段时间,以胖为美。即使在今天,在许多国家,肥胖也会与富有联系在一起。然而,在大多数地方,瘦即意味着成功。随着时间的推移,这种情况可能也会改变,不变的

for a person's health. Fat is not healthy; it makes your body work harder by putting more stress on every interactive part of your being. Most importantly, every extra pound you carry adds an extra load on your heart which increases your risk of cardiac arrest.

Here's a definitive example of carrying around extra weight. When I was in college, a roommate asked me to grab two five-pound dumbbells and walk around for one hour. I was young and in decent shape, so it wasn't all that strenuous. He then had me try ten-pound dumbbells and do the same. This time, it did not take long for me to run out of breath and find myself exhausted. The extra weight I toted was exactly the same as lugging around an extra twenty pounds of fat! Your body will suffer from being overweight and will take its toll on your overall well-being. As I said before, *The Baby Food Diet* will help you live longer by removing toxins from your body while lowering your mass.

第五章
心态和其他有趣的想法

是,保持体形对一个人的健康更有好处。肥胖根本不健康,因其会对身体的每一个互动部分施加更多的压力,而让身体更辛苦。最重要的是,每增加一磅体重,心脏就会增加额外的负荷,从而增加心脏骤停的风险。

这是一个关于超重的典型例子。当我上大学的时候,一个室友让我拿两个5磅重的哑铃,走一个小时。当时我很年轻,身材也很好,所以并不怎么费力。然后,他让我试了试10磅重的哑铃,还是走一个小时。这回,没过多久,我就喘不过气来了,并且发现自己筋疲力尽。我拿在手里的额外重量正像是身上拖着的额外的20磅脂肪一样!你的身体会因为超重而受苦,并对整体健康造成损害。就像我之前说的,**婴儿食品节食法**可以在降低体重的同时排出体内的毒素,并由此让你活得更长久。

CHAPTER 6

Benefits of The Baby Food Diet

第六章

婴儿食品节食法的益处

CHAPTER 6
Benefits of The Baby Food Diet

To this juncture, I have touched upon some of the positive benefits of the diet. Below is a more comprehensive list of all the productive reasons to join *The Baby Food Diet* revolution:

1. Being thinner will make you more attractive to others. This will not only help your professional career, but your personal life as well. First impressions and appearance are critical in today's business world and on the social scene. It can even reignite the fire in your marriage.

2. Being overweight is stressful, creates intense worry, and can lead to crippling self-consciousness in public situations. Once you are thin, there is one less thing to worry about. Stress is cumulative and reminds me of the old saying, "The last straw that broke the camel's back." The more stress that can be removed from your life, the less burden on your health.

3. Your attitude will improve because of the pleasure you derive from your svelte physique. You will develop more confidence which is paramount along the highway to success. A more positive outlook will enhance the way you approach and resolve issues. A good attitude is always a winning formula.

4. You will become more energetic. This newfound vigor will lead you to be more focused, more productive. You will be more prone to exercise. Most people do not

第六章
婴儿食品节食法的益处

到目前为止，我已经提到了节食的一些益处。下面是一份更全面的清单，列出了所有富有成效的加入**婴儿食品节食法**革命之原因：

1. 身材好能使你在他人眼中更有吸引力。这不仅对职业生涯有助益，对个人生活也有帮助。在当今商务世界和社交场合，第一印象和外表至关重要，它甚至可以重新点燃你的婚姻之火。

2. 超重会带来压力，产生强烈的担忧，并会导致在公共场合严重的自我意识缺失。一旦你瘦下来，就少了一件担心的事。压力是日积月累形成的，让我想起一句老话：压垮骆驼的最后一根稻草。你将生活中的压力甩掉得越多，你的健康负担就得越少。

3. 你的状态会有所改善。因为你从苗条的身材中获得了快乐，你会越发自信，而在通往成功的道路上，自信是至关重要的。一个更加积极的形象会增强你处理和解决问题的方式。良好的状态永远是成功的秘诀。

4. 你会变得更加精力充沛。这一新兴的活力会让你更

work out because they are too tired. With this abundance of natural energy, you will be much more active in all areas of your life.

5. A thin person is more likely to stay with an exercise program. When I got heavy, I was constantly injuring myself and eventually stopped going to the gym for this reason. I quit playing my favorite sport, which is basketball, as the weight caused issues with my Achilles tendon. Now that I am thin, I have resumed playing basketball and have even taken up bicycle riding again, and it is lots of fun.

6. Being thin will not only give you more energy to exercise, it will compel you to live a more active lifestyle on all levels. I remember being at the beach once and was too embarrassed to take off my shirt and go swimming. Hiking was out of the question because an obese person has a difficult time trekking up hills. Even when riding a bicycle, overweight individuals worry about balance and falling, not to mention the small seats are very uncomfortable if your rear end is plentiful.

7. You will be more receptive to having your picture taken and viewed by others. Many overweight people are embarrassed by the way they look and detest being photographed. This is unfortunate for many reasons. For example, I have a habit of taking pictures of people I meet. This is a good method for me to remember new

第六章
婴儿食品节食法的益处

专注,更有效率,你会更乐于运动。大多数人不愿健身是因为他们太累了,有了这些丰富的自然能量,你会在生活的各个方面更加活跃。

5. 身材好的人更有可能坚持健身计划。当我变胖的时候,经常受伤,最终因为这个原因我不再去健身房,放弃了最喜欢的运动——篮球,因为体重的原因引发了跟腱问题。现在我瘦了,又开始打篮球了,而且又开始骑自行车了,真的非常有趣。

6. 瘦下来不仅让你有更多的能量去运动,还会迫使你在各个层面上拥有一种更积极的生活方式。记得有一次,在海滩上,我甚至不好意思脱掉衬衫去游泳。徒步旅行就更不可能,因为肥胖的人爬山会很困难。即使是骑自行车,超重的人也会担心平衡问题导致摔倒,更不用说如果臀部太大,那么小的车座肯定会令人很不舒服。

7. 你会更容易接受别人为你拍照,并让人欣赏你的照片。许多超重的人对自己的外表感到尴尬,讨厌被拍照。这很不幸,原因有很多。例如,我有给见面的人拍照的习惯。对我来说,这是一个记住新相识的朋友的好

acquaintances which is significant in building relationships, both personal and business. Once you are thin, a camera will not make you shudder!

8. Your loved ones will stop bothering you and worrying about your health. When I was heavy, my family and friends would harp on me about the extra weight being detrimental to my health. They were always giving me advice on this and that concerning my fitness. Now they never worry, which in turn makes me happy.

9. Being fit and trim sets a great example for your children and friends. Have you ever noticed most fat people have fat children? Sometimes this is because of hereditary reasons, but most often it is because the children devour the same high calorie diet as the overweight parent. Being a healthy role model is a powerful image.

10. You will also save money for the inherent inconveniences associated with being overweight. Instead of always having to fly business class or first class to accommodate your bulk, you can comfortably fly coach. Even if you can't afford to fly the premium seats, how much more comfortable will the coach seat be once you are thin?

11. The comfort you will enjoy in everyday life is a huge benefit. The world is not accommodating for fat people. Climbing in and out of cars, bending over to pick things up, and finding clothes that properly fit are

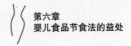
第六章
婴儿食品节食法的益处

方法，对建立人际关系和商业关系都很重要。一旦你瘦下来，相机就不会再让你发抖！

8.你的亲人不会再来烦你并担心你的健康。当我很胖的时候，我的家人和朋友会喋喋不休地诉说超重对健康有害，他们总是提一些关于身体健康的建议。现在，他们再也不用担心了，这反过来使我很高兴。

9.保持健康和匀称的身材会为子女和朋友树立很好的榜样。你有没有注意到，大多数胖人他们的孩子也都很胖？有时这是由于遗传的原因，但大多数情况下，是因为孩子们和超重的父母一样，吃同样高热量的食物。健康的形象是强大的榜样。

10.因为超重带来的固有不便消失后，你会因此而省钱。你可以舒适地乘坐经济舱，而不必总是乘坐商务舱或头等舱以适应体形。即使你坐不起头等舱，一旦你瘦下来，经济舱的座位会有多舒服？

11.你在日常生活中能更加舒适。这个世界不适合胖子，上下汽车、弯腰捡东西、找到合身的衣服，都是每天的挑战。生活的方方面面都受到身体胖瘦的影响。做

daily challenges. Every aspect of life is influenced by the size of your body. In business, we often caution staying within the box to effectively achieve results. I would also say: stay within the limits of the weight box to achieve your desired results and appreciate the world from a better place.

12. Above we talked about some of the countless health benefits of losing extra weight. A few more bonuses include the possibility of lowering your blood pressure, preventing heart disease, eliminating shortness of breath, improved flexibility, reduced stress on your joints, and so on. *The Baby Food Diet* is your membership card to the exclusive thinner, healthier and longer life club.

13. How about looking in a mirror? Even if you try to avoid mirrors, it is impossible not to catch a glimpse of yourself in a window reflection or your shadow on a wall or sidewalk. Imagine you are in a meeting and excuse yourself to go to the bathroom. While washing your hands, you see your reflection in the mirror. Does your image give you confidence or does it deflate your attitude? If you are not satisfied, the answer once again is to shed the pounds which will develop better self-confidence and have you returning to the meeting with a bounce in your step.

14. You will sleep better at night if thin. This is a well-known, indisputable fact. Many doctors believe that

第六章
婴儿食品节食法的益处

生意时,我们经常告诫自己,要守住条条框框,才能有效地实现目标。我还想说的是,为了达到你想要的结果,保持在体重的条框之内,从更好的角度欣赏这个世界。

12. 以上我们讨论了减肥所带来的很多健康益处。还有一些额外的好处,包括降低血压、预防心脏病、消除呼吸短促、提高灵活性、减轻关节压力等等方面的可能性。**婴儿食品节食法**是你专属的瘦身、健康、长寿俱乐部的会员卡。

13. 想照照镜子吗?即使你试图避开镜子,也躲不开从窗户玻璃的反光中,或者在墙上以及人行道上瞥见自己的影子。假设你正在开会,得空去个洗手间,当你洗手的时候,也会看到镜子里的自己。你的形象是让你拥有信心,还是会让你泄气?如果你不满意,那么再说一次,解决办法就是减肥,这会让你更加自信,当你重新回到会议中时,会精神抖擞。

14. 如果你瘦下来,晚上会睡得更好。这是众所周知的不争事实。许多医生认为,超重的人比体瘦的人更容易患睡眠呼吸暂停症。如果你睡得很香,醒来时就会有更多的

overweight people are more subject to sleep apnea than thin people. If you sleep more soundly, then you will wake up with more energy which will last throughout the entire day. A person who is well rested makes better and more rational decisions. Can better sleep help you with depression? Most doctors believe it can. There are many books on the importance of sleep. People all over the world are hooked on sleeping pills. If you are among them, after a few months on *The Baby Food Diet*, you will no longer need them. It is fact, a good night's sleep is one of the keys to a happy life.

15. Being on *The Baby Food Diet* will make you proud. Your self-esteem will soar knowing you are taking control of your life and are committed to improving who you are. It is time to quit with the excuses and take action that will instill a sense of pride and accomplishment. You will feel a glorious sense of achievement.

16. Generally, thin people sweat less than fat people due to their better overall shape. In a meeting or on a date, sweating is a cause of internal alarm. It is a confidence buster when you are the only one at the table dripping with sweat. Lose the weight, lose the sweat.

17. Sometimes being overweight is likened to being wrapped in chains or trapped in prison because of the associated limitations. There are many occasions where extra pounds can restrict the richness of life experiences due to intense self-consciousness. It can be an excuse

第六章
婴儿食品节食法的益处

能量，可以持续一整天。一个充分休息的人会做出更好、更理性的决定。更好的睡眠会有助于克服抑郁症吗？大多数医生认为可以。有很多关于睡眠重要性的书籍。世界上很多人都服用安眠药上瘾，如果你是其中一员，经过几个月的**婴儿食品节食法**，你就不再需要安眠药了。事实上，晚上睡个好觉是幸福生活的关键之一。

15. 尝试**婴儿食品节食法**会让你感到自豪。当意识到能够掌控自己的生活，并致力于改善自身时，你的自尊感会飙升。不要再找借口，是时候采取行动灌输自豪感和成就感了，你会感到一种辉煌的成就感。

16. 一般来说，瘦人因其整体身材较好，要比胖人出汗少。在开会或约会的时候，出汗会引起内心恐慌。当你是在座者中唯一汗流浃背的人时，自信心就会受到打击。只有减去体重，才能不再流汗。

17. 有时候，因为与受限制相关，超重的人就好像被锁链捆绑起来，或者被困在监狱里一样。在很多情况下，由于强烈的自我意识，超重会限制生活体验的丰富性。这也会是错过一个好机会或一些乐趣的原因。现在，由于减轻了体重，再没什么能阻止我去接触新

to miss out on a good opportunity or some fun. Today, thanks to my weight loss nothing holds me back from new and exciting encounters.

18. *The Baby Food Diet* will enable you to stop feeling jealous or inadequate when around your thin friends. The fact is: being overweight is something you can fix. The choice is solely yours. When you see a thin person, it is easy to envy them for the wrong reason. Are you believing that they are better than you? Are you upset because they look more attractive and have more discipline than you? Discipline is merely training, and that's all it takes to lose weight. You are as good as anyone else and maybe better, so why not find out!

19. Since I became thin, I noticed that I drink less alcohol and generally engage in fewer things that harm my body. I believe this is a subconscious phenomenon that just happens naturally when anyone takes control of their weight. Deep down you feel healthy, and there is a newfound sense of willingness to be healthy. Don't worry, you will still be a fun person, just a little more health-conscious than you used to be when heavier.

20. Everything in life will become easier, more comfortable. You will be able to find clothes that fit without always going to the tailor. You will be more agile and move around more easily. You will sleep better due to an increased comfort level.

第六章
婴儿食品节食法的益处

鲜的、令人兴奋的事物了。

18. 当你和体形好的朋友在一起时，**婴儿食品节食法**让你再也不会感到嫉妒或不足。事实是：超重是有办法解决的，选择权完全在你。当你看到一个纤瘦的人时，很容易出于错误的原因而嫉妒。你是否相信他们比你更优秀？你是否因为他们看起来比你更有吸引力、比你更自律而感到沮丧？自律只要经过训练就可以达到，而这也正是减肥所需要的全部。你和其他人一样好，也许更好，所以为什么不去发现自己的好！

19. 自从瘦身以后，我注意到自己喝酒少了，慢慢也很少做伤害自己身体的事情了，我相信这是一种下意识的现象，当人控制自己的体重时，这种现象就会自然而然地发生。在内心深处，你觉得自己很健康，并且有一种乐于健康的新感觉。别担心，你仍然会是那个有趣的人，只是比以前胖的时候更有健康意识而已。

20. 生活中的一切都会变得更容易，更舒适。你不需要经常去裁缝那里就能找到合身的衣服。你会变得更加敏捷，行动自如。由于舒适水平的提高，你会睡得更香。

CHAPTER 7

Protein and Baby Food

第七章

蛋白质和婴儿食品

CHAPTER 7
Protein and Baby Food

I am often asked what baby food is best when on this diet. First, it comes down to your individual taste. You must experiment with different baby foods and brands. Reading the labels to understand the ingredients is always a good idea. You will find that normal store branded baby foods are similar because a baby's taste is not yet fully developed, and their digestion systems are very sensitive.

In the future www.thebabyfooddiet.com will offer its own branded versions of both baby food and proteins. Of course, these will be your best option, as they will have ingredients that babies cannot digest. Vitamins, enhanced flavors, and other healthy additives will be part of our formulas and recipes. Until then, store brands are an excellent source.

For our in-house brand, which is now in the final phases of research and development, you will experience the optimum dieting results. These internally developed baby foods are really a type of superfood and as close to the perfect nourishment you will ever find. Baby food plus liquid and powder protein products are now in the final stages of being created.

When I was in high school, I participated in amateur wrestling. Wrestlers are categorized by weight class, so they are always trying to cut down on their weight. Thus, I was first introduced to protein powder

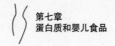

第七章
蛋白质和婴儿食品

在这种节食法中,经常有人问我哪种婴儿食品是最好的。首先,这取决于个人口味。需要尝试不同的婴儿食品和品牌。阅读标签以了解食物成分从来都是个好主意。你会发现,普通商店里各品牌的婴儿食品都很相似,因为婴儿的味蕾还没有完全开发,而他们的消化系统又非常敏感。

今后我们会在 www.thebabyfooddiet.com 网站上,推出自有品牌的婴儿食品和蛋白质。当然,这些将是你最好的选择,因为其中会包含婴儿无法消化的成分。维生素,增强型口味和其他健康添加剂将成为我们配方和食谱的一部分。而在此之前,商店品牌则是一个很好的来源。

对于我们的自有品牌,目前正处于研发的最后阶段,你将体验到最佳的节食效果。这些自主开发的婴儿食品确实是一种超级食品,几乎是你能找到的最完美的营养品。婴儿食品加上液体和粉状蛋白质产品目前正处于研制的最后阶段。

我上高中的时候,曾参加过业余摔跤活动。摔跤运动员按体重划分级别,所以他们总是试图减体重。也因

which is an important ingredient of *The Baby Food Diet*. To start with, all proteins are not the same. There are many kinds of powders including whey, soy, pea, and hemp seed protein. Another important consideration is that people have different palates, so try more than one of these protein powders to find the one that tastes best to you. Before purchasing any of them, read the ingredients and labels to make sure you understand the number of calories and protein per serving.

While on the Baby Food Fast, drink protein once or twice each day. I use a blender to avoid any clumps in the protein. I have found even the best, most expensive protein powders have this issue of clumping, despite their advertising claims. When traveling, I always bring a small electric blender and my protein supplement. After mixing the protein, I simply drink it directly from the container which is very efficient and easy to pack.

I only use water to mix with the protein. I never use milk. At times, I will add a few blueberries, strawberries, bananas or other fruits to the concoction. I suggest using small portions, maybe 15 blueberries, 3 strawberries or up to half of a banana. Keeping track of how much fruit you add ensures that there aren't too many extra calories in your shake. If you do use fruit, limit it to only once per day. If you are doing two shakes each day, then one of them should not have any fruit.

第七章
蛋白质和婴儿食品

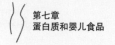

此，我第一次接触到蛋白粉，这是**婴儿食品节食法**中的重要成分。首先，并非所有的蛋白质都是一样的。有许多种类的蛋白粉，包括乳清、大豆、豌豆和火麻仁蛋白。另一个重要的原因是，人的口味各不相同，所以要尝试多种蛋白粉来找到你最喜欢的那种。在购买任何一种之前，要阅读成分和标签，以确保清楚每份中的卡路里和蛋白质的含量。

在婴儿食品禁食期时，每天喝1～2次蛋白奶昔。我用搅拌器来避免蛋白粉结块，我发现即使是最好的、最贵的蛋白粉也会出现这种结块现象，尽管他们在广告中声称不会。旅行时，我总是带一个小型电动搅拌器以及蛋白质营养补充剂。在搅拌完蛋白粉后，我直接从搅拌器杯子里喝，这样做既有效又容易携带。

我只用水与蛋白粉搅拌，从来不用牛奶。有时，我会在搅拌物中加入一些蓝莓、草莓、香蕉或其他水果。我建议加入小份，比如15颗蓝莓、3颗草莓或者最多半根香蕉。记录下你加了多少水果，确保当中没有太多额外的卡路里。如果你确实要加水果，要限定每天只加一次。如果你每天要搅拌两杯蛋白奶昔，那么

Really, I only add fruit about once a week and just for a bit of variety.

You can also add ice to the blender or cold water and the taste is even better, more refreshing. I never use ice, but I do use cold water. When there is no cold water available, I just use room temperature water, and it still tastes great. In China, people believe cold water is bad for digestion so there is nothing wrong with using even warm water. There is merit in this rationale, so if you want the very best results use room temperature or warm water. For me, chilled water is preferential because I just love the refreshing taste of a nice cold protein shake.

Protein shakes can be continued in all phases of *The Baby Food Diet*. However, it is not necessary to continue drinking protein shakes to maintain your target weight on the Baby Food Maintenance Diet. For me, even when on the maintenance diet, I continue daily to consume both protein shakes and baby food. You can make it a personal choice whether to continue the protein once you reach the point of maintenance.

There are times in all our lives when upcoming events stimulate an urgency to lose weight. Maybe a gathering that you had previously not planned on attending suddenly becomes alluring because old friends will be there. Maybe it is your high school reunion or something similar. Whatever it is, *The Baby Food Diet* is

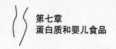
第七章
蛋白质和婴儿食品

其中一杯不应该有任何水果。真的,我一周只加一次水果,只是为了增加一点多样性。

你也可以在搅拌器里加冰或者凉水,味道会更好、更清爽。我从不加冰,但我会用凉水。当没有凉水可用时,我就用白开水,味道仍然很好。在中国,人们认为凉水不利于消化,所以即使用温水也没问题。这一理论自有其好处,所以如果你想要最好的结果,请使用白开水或温水。对我来说,冷水是首选,因为我喜欢清凉爽口的蛋白奶昔。

蛋白奶昔可以在**婴儿食品节食法**的所有阶段持续饮用。然而,在婴儿食品维持期没有必要继续喝蛋白奶昔来保持目标体重。对我来说,即使是在维持期,我仍然每天食用蛋白奶昔和婴儿食品。一旦达到维持期,你可以自行决定是否继续食用蛋白粉。

在生活中我们总有一些时候,会有即将发生的事情促使我们迫切地减肥。也许是你之前没想参加的聚会,由于老朋友会出现而突然变得很有吸引力;也许是高中同学聚会或者类似的集会。不管是哪种,**婴儿食品节**

your answer to losing weight rapidly, besides being a long-term solution. If you need to lose 20 or 30 pounds in a hurry, then try *The Baby Food Diet* followed by the Baby Food fast, and you will achieve super-fast results.

The Baby Food Fast can be continued for as long as fourteen to twenty-one days and is as simple as can be. With the Baby Food Fast, you will only be eating two baby foods a day and drinking one or two protein shakes. On the days when you cannot stand the regimen anymore, you can enjoy an apple or an extra baby food. It is that easy.

Once you get to the fourteenth day, you should revert to the Baby Food Reduction Diet and keep going for a while. It is ok to go back on the Baby Food Fast after a week on the Baby Food Reduction Diet. This ensures you to stay healthy and strong, but also lose weight at a fast and steady pace. *The Baby Food Diet* is suitable for all people, and as such really is indeed the perfect diet.

The Baby Food Fast works amazingly well and the results will last as long as you continue to follow the Baby Food Maintenance Diet. For sustainability, it is important that eating has a fun element attached. Even the Baby Food Fast is fun and enjoyable for me. However, it does take some of fun out of eating on a social occasion. Imagine going to a restaurant and only

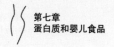

第七章
蛋白质和婴儿食品

食法不只是一个长期的解决方案,也是快速减肥的办法。如果你需要在短时间内减掉20或30磅的体重,那么试试**婴儿食品节食法**,按照婴儿食品禁食期的办法来做,你就会达到超高速的效果。

婴儿食品禁食期最多可以持续14~21天,而且非常简单。在婴儿食品禁食期,你每天只能吃两份婴儿食品,喝1~2份蛋白奶昔。在忍无可忍的日子里,你可以享受一个苹果或额外的一份婴儿食品。就这么简单。

一旦到了第14天,你应该回到婴儿食品减食期,并坚持一段时间。遵循婴儿食品减食期一周后,再重新回到婴儿食品禁食期是没有问题的。这不仅能确保你保持健康和强壮,还能快速稳定地减肥。**婴儿食品节食法**适合所有人群,因此确实称得上完美的节食法。

婴儿食品禁食期的效果好到惊人,只要你之后继续遵循婴儿食品维持期,这一成果将会持续。为了可持续性,饮食中有趣的元素很重要。对我来讲,甚至婴儿食品禁食期也是有趣和愉快的。然而,在社交场合吃这些东西确实会失去一些乐趣。想象一下,要是到一家

having one portion of baby food?

The Baby Food Diet, especially the Baby Food Maintenance Diet phase, gives you lots of flexibility. Of course, I even advise on the Baby Food Maintenance Diet to watch your portions. Maybe once per week, it is okay to totally blow it. Maybe when you are on a two-week vacation, you can forget the diet altogether and just relax. As long as you realize that when you return to the diet, there will be a heavy price to pay. If you have previously committed to *The Baby Food Diet*, then you already know that a viable solution exists. Two weeks is not so long, and a rich life is best enjoyed with a pinch of flexibility. *The Baby Food Diet* allows for this flexibility and a few culinary indiscretions.

The good news is if you stop the diet, say after six months of abandoning it, *The Baby Food Diet* will call you back. The diet makes your energy levels so high that you will miss that great feeling that you experienced while dieting. It normally takes anywhere from 6 to 8 weeks of being on the diet to get to this point of missing its constructive benefits. Staying or returning to the diet will make you feel younger, more special, and best of all, you will be lean and healthy.

While this diet makes it easy to lose the weight, many of life's challenges and enticements make it just as easy to quit. Honestly, how hard can it be to open a

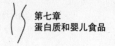

第七章
蛋白质和婴儿食品

餐厅去吃饭,而你却只能吃一份婴儿食品?

婴儿食品节食法,尤其是在婴儿食品维持期,可以给你很大的灵活性。当然,我还是建议即使在婴儿食品维持期的时候也要注意分量控制。也许一周可以有一次,完全不按规矩来也没关系。也可能当你在两周的假期里,完全忘记节食这件事,只管放松一下。只要你意识到,当你重新回到节食上时,将会付出沉重的代价。如果你以前坚持过**婴儿食品节食法**,那么你已经知道有一个可行的解决方案。两周时间并不长,只要有一点灵活性,就能享受丰富多彩的生活。**婴儿食品节食法**允许这种灵活性和一些烹饪上的小任性。

好消息是,如果你停止节食,比如说放弃6个月后,**婴儿食品节食法**会召唤你回来。节食让你的能量水平如此之高,以至于你会想念节食时那种美妙的感觉。正常情况下,需要坚持节食6~8周的时间,才能达到想念这些好处的程度。坚持或恢复节食会让你感觉更年轻、更特别,最重要的是,你会变得瘦而健康。

虽然这种节食法很容易减肥,但生活中的诸多挑战和

container of baby food and drink protein shakes? However, once you have lost the weight new obstacles arise, and it is essential that you learn to read the ingredient labels. Check the number of calories and nutritional value of what you are about to consume. Avoid highly processed foods with lots of chemicals in them. Put an app on your phone to easily check any food you are considering digesting. This only takes seconds.

As explained, there is failsafe method of keeping off the weight once it has been lost, by simply weighing yourself once a month. If you have experienced more than a 5-pound gain you immediately return to one of the more extreme versions of *The Baby Food Diet*. I question, why would you want to continually put yourself in this position? You can avoid going back to the Baby Food Fast and Baby Food Reduction Diets by taking the time to make informed food choices while on the Baby Food Maintenance Diet. It is much more fun to be on the maintenance part of *The Baby Food Diet* than the other two options.

Again, read food labels at markets and use an app at restaurants to understand exactly what you are ordering and how it will affect your weight. This does not take much time and will keep you healthier.

第七章
蛋白质和婴儿食品

诱惑也会让放弃节食变得很容易。老实说,打开一瓶婴儿食品并喝点蛋白奶昔能有多难?然而,一旦你减掉了体重,新的障碍就会出现,你必须学会阅读成分标签,检查即将摄入食物的卡路里和营养价值。避免含有大量化学物质的高度加工食品。在手机上安装一个应用程序,可以方便地查看你正在考虑进食的食物。这只需要几秒钟。

如前所述,有一种方法可以在一旦减肥成功后继续保持体重,那就是每月只称一次体重。如果你的体重增加超过 5 磅,马上回到**婴儿食品节食法**的极端版本。我想问的是,为什么要一直把自己推到这个位置上呢?只要在婴儿食品维持期的时候,花时间做出明智的进食选择,你完全可以避免回到婴儿食品禁食期和婴儿食品减食期。比起**婴儿食品节食法**中另两个阶段来讲,婴儿食品维持期要有趣得多。

此外,在商场里阅读食品标签,在餐馆里使用应用程序来了解所点食物的情况以及它将如何影响你的体重。这不会花太多时间,而且会让你更健康。

CHAPTER 8

**Psychological Part of
the Diet**

第八章

节食中的
心理因素

CHAPTER 8
Psychological Part of the Diet

Now I will emphasize the psychological part of the diet. There are many diets, and granted, if you follow them, you will lose weight. Getting started is the toughest part. Sticking to any diet is second toughest. Success is all about discipline and mind over matter. The beauty of *The Baby Food Diet* is that once started, it is so easy to stay with.

So, how do you overcome the obstacle of self-defeat? Worry not! There is a way, and it involves the retraining of your subconscious mind. There are many books on this subject, and I advise you buy one and learn more. I will do my best to teach you the basics here. But really, it is best to consult one of these well researched books. The secret is something called "Self-Talk", which is telling yourself positive things and retraining your way of thinking.

For Self-Talk to work, you must diligently practice it for a few minutes each day. The first step is to write down the changes and goals you set in the form of affirmations. Then repeat the declarations out loud to yourself every day to achieve the best results. I recommend doing this twice a day. When you wake up and again when you go to bed it works best for me. In addition, jot down three of your affirmations on a separate piece of paper to consult during the day.

There are many ways you can use affirmations. I

第八章
节食中的心理因素

现在我要强调节食中的心理因素。有很多种节食方法,当然如果你遵循它们,体重肯定会减轻。着手开始是最难的部分,能坚持下去则是任何节食法中的第二个难题,成功源于自律和精神,这胜于物质。**婴儿食品节食法**的美妙之处在于,一旦开始,就很容易坚持下去。

那么,你怎样才能克服自我挫败的障碍呢?不要担心!有一个方法,它涉及潜意识的再训练。关于这方面的书籍有很多,我建议你买一本,多学一些。在此我将尽最大努力教给你基本知识。但实际上,最好是翻阅一本在这些方面研究得比较充分的书籍。秘诀就是所谓的"自言自语",告诉自己一些积极正向的事情,并重新训练自己的思维方式。

要想让自言自语发挥作用,你必须每天勤勉地练习几分钟。第一步是以肯定的形式写下你所设定的改变和目标,然后每天对自己大声重复这些宣言,以达到最好的效果。我建议每天做两次:一次是早上刚醒来的时候,另一次是晚上准备睡觉之前。这对我特别有效。此外,在另外一张纸上写下你的三个自我肯定,以便在白天的时候翻看。

你可以用很多方法来肯定自己。我有时会把写下的东

sometimes record what I wrote down, then listen to it in the car or at the gym. As a last step, I post these affirmations in places where I can see them, like on the refrigerator or bathroom mirror.But all you really need to do is recite them aloud once a day to benefit. However, realize that every reinforcement creates faster results.

Self-Talk works for one reason: "You are what you think." Your thoughts power your subconscious mind. What your subconscious mind thinks determines your actions. By writing down what you want in life and saying it out loud, you are retraining your intuition. You can do this for all parts of your life. Start with *The Baby Food Diet* and your target goals, but as time passes expand Self-Talk to other parts of your life.

Trust me when I tell you it works. It may take a little time, but you will realize enormous positive changes through Self-Talk. Like all things, consistency and daily practice are the keys to success. I recommend a further reinforcement step; rewrite your goals every two weeks, even if they remain unchanged. The act of writing down goals will lead to a faster result.

Here are some examples of Self-Talk:

I will change my body by following The Baby Food Diet.
I will live a healthier life by always making healthful choices.

第八章
节食中的心理因素

西录音,然后在车里或健身房里听。作为最后一步,我会把这些肯定的话贴在所能看到的地方,比如冰箱门上或浴室镜子上。但你真正需要做的是每天大声背诵一遍,以便从中受益。然而,要意识到每巩固强化一次都会产生更快的结果。

自言自语之所以奏效的原因是:"其思即其人。"你的思想支配着你的潜意识;你潜意识的想法决定你的行动。通过写下在生活中想要什么并大声说出来,你就是在重新训练自己的直觉。你可以在自己生活中的所有方面这样做。**先从婴儿食品节食法**和你的目标开始,然后随着时间的推移,将自言自语扩大到生活中的其他方面。

当我告诉你它能起作用的时候要相信我。这可能需要一点时间,但你会通过自言自语实现巨大的积极的变化。像所有事情一样,坚持不懈和每天练习是成功的关键。我建议采取进一步的巩固措施:每两周重新写一次你的目标,即使内容没有变化。写下目标的行为会导致更快地实现目标。

下面是一些自言自语的例子:

CHAPTER 8
Psychological Part of the Diet

I do not care what others say, I will stick with The Baby Food Diet

Everyday my weight is lower, and I am healthier.

I love the way I look.

I am attractive and carry myself well.

I am a successful and happy person.

I exercise regularly and enjoy exercising.

Even on busy days, I do things to stay in shape, like walking upstairs instead of taking the elevator.

At all times, I have a winning attitude.

People like me for who I am.

I am on this diet for myself as I want to be the best I can be.

I am committed to this diet forever.

I always make wise choices in life.

I listen, hear and learn.

I appreciate others and I am thankful for being me.

I am a giving and caring person.

This diet is wonderful, and I hope others can join me and get thin and healthy.

Today is the best day of my life.

I am a success in every way that can be defined.

When I want something, I work smart and hard to obtain what I desire.

I set goals, and I always achieve my goals.

I am a very balanced person.

I make a good friend and family member.

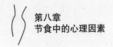

第八章
节食中的心理因素

通过遵循婴儿食品节食法,我会改变自己的身体。

我会通过做健康的选择来过更健康的生活。

我不在乎别人怎么说,我会坚持婴儿食品节食法。

我的体重每天都在减,我也更加健康。

我喜欢自己的样子。

我很有魅力,举止得体。

我是一个成功和快乐的人。

我经常运动,并喜欢运动。

即使在忙碌的日子里,我也会做一些事情来保持体形,比如爬楼梯而不是坐电梯。

任何时候,我都有一种必胜的信念。

人们喜欢真实的我。

我为自己节食,因为我想做最好的自己。

我会永远坚持这一节食方法。

我在生活中总是做出明智的选择。

我会倾听,听进,学到。

我感激他人,也感谢能成为我自己。

我是一个乐于奉献和关心他人之人。

这种节食方法很好,我希望其他人能加入我,变得苗条而健康。

今天是我一生中最美好的一天。

我在各个方面都是成功的。

I am a great communicator.
I am an example of success and happiness.
This diet is amazing
My willpower is strong, and it is easy for me to stay on the diet.

These are some examples of affirmations that you can write and then say out loud. You can build your own list. Maybe you can add some new items and remove others. Don't be shy because this only works if you read your Self-Talk avowals out loud to create a new you.

I have been using this Self-Talk technique for 25 years and credit it for much of my transformational success. One thing I have learned is to take it seriously. Do not stop once you attain your goals. Instead, add new items and continue to develop yourself. To me, life is about balance and growing. To learn new things and never be a "know-it-all" are traits of truly successful and content people.

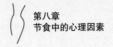
第八章
节食中的心理因素

当我想要什么东西的时候,我会聪明、努力地工作,去得到我想要的。

我设定目标,也总能实现目标。

我是一个非常平衡的人。

我是一个好朋友和好家人。

我是一个最好的沟通者。

我是成功和幸福的典范。

这种节食方法真的很棒!

我的意志力很强,对我来说坚持节食很容易。

这些都是自我肯定的例子,可以写下来,然后大声说出来。你可以列出自己的清单,也可以添加一些新项目和删除原有的一些。不要害羞,因为只有当你大声说出这些自言自语时,它才会更好地起作用,由此创造一个全新的自己。

我使用这种自言自语的方法已经有25年了,我的转型成功很大程度上归功于此。我从中学到的是要认真对待此事,即使你达到了目标也不要停止。相反,可以增加新项目,继续开发自己。对我来说,生活就是平衡和成长。永远不要成为一个"万事通",要学习新鲜事物,这才是真正志得意满人士的特征。

CHAPTER 9

Exercise

第九章

运动

CHAPTER 9
Exercise

Let's now talk about exercise, and why I think it is important. Everyone has a different body, so no one program fits all people. There are several factors that influence the kind of exercise regime you should use. These include age, body type, medical conditions as well as time and money. Some people claim that they have very little free time to designate for working out. To me this excuse is ridiculous as allocating time for vital needs should always be done. Other people have an abundance of free time and still don't appropriate any physical effort into their lives. Your age and overall health condition will dictate the type of activities you should undertake as part of your exercise regimen.

The purpose of exercise is not to lose weight; it is to keep your heart healthy and build up strength. Most of the weight loss will come from your food consumption habits. The fastest way to lose weight is to modify your diet. *The Baby Food Diet* will be the reason you lose weight, not merely working out. This said, a solid exercise program can help you realize a more toned and balanced appearance. In addition, it is widely known that even the most moderate of a physical program can have huge health benefits. Exercise can also give you some quiet time to contemplate or meditate.

For some people, exercise could be regular sessions of yoga or planking. For others, it may be a brisk walk or dance class with others. I enjoy going to the gym.

第九章
运动

现在我们来谈谈运动,以及为什么我认为它很重要。每个人的身体都不同,所以没有一项运动可以适合所有人。有几个因素会影响应该采用哪些锻炼方式,这些因素包括年龄、体形、身体状况以及时间和消费水平。有些人声称他们很少有富余时间专门去锻炼。对我来说,这个借口是荒谬的,因为把时间分配给重要的需求历来都是理所应当的。另一些人则有大把富余的时间,但他们仍然在生活中不做任何适当的健身运动。你的年龄和整体健康状况将决定你应该进行什么样的活动,这是锻炼计划的一部分。

运动的目的不是为了减肥,而是要保持心脏健康和增加力量。减肥的大部分都是来自饮食习惯,最快的减肥方法是改变自身的饮食习惯。**婴儿食品节食法**会成为你减肥的绝佳方法,而不是仅仅依靠锻炼。也就是说,一个扎实的锻炼计划可以帮助你呈现更加健美和匀称的外表。此外,众所周知,即使是最平和的体育锻炼也能带来巨大的健康益处。健身也能给你提供一些安静的时间来思考或冥想。

对一些人来说,锻炼可以是定期的瑜伽或平板支撑。对其他人来讲,可能是一次轻快的散步或与别人一起上舞

CHAPTER 9
Exercise

Swimming is an excellent choice as it develops lean muscles and puts less stress on your joints and ligaments. I used to jog, but the pounding of my legs eventually started to cause knee and back problems. I switched to cycling, weight training and other cardio workouts.

When it comes to exercise, choose activities that suit your needs and tastes. Make sure you balance out cardio with resistance training. You can join a class or two with an instructor to ensure that you are doing it properly which is a huge benefit. Remember, even if you do not belong to a gym, there are many free activities that you can perform at home or in a park. Check online for innovative exercise programs as there are thousands available at no cost. The key is to be consistent and increase the intensity or session length over time, and you'll experience improvement.

Exercising with a group is a good way to meet new people. There are hiking and biking clubs, martial arts schools, skating groups, dance troupes, and the list goes on. If you seek new friends, physical activities provide a great way to encounter diverse individuals. If you are a person who finds it easier to exercise with others, then join a group. There are apps on your phone that can help you find such meet-ups or you can do internet searches. If you see a group in action, don't be shy, walk up and ask to join. If you end up not liking that assembly, you can remain

第九章
运动

蹈课。我喜欢去健身房。游泳是一项很好的选择,因为它能增肌,减少关节和韧带的压力。我习惯于慢跑,但腿部的撞击最终导致我的膝盖和背部开始出现问题,我又改成骑自行车、举重和其他有氧运动。

说到健身,要选择适合自身需求和品味的活动,确保你在有氧运动和力量训练之间取得平衡。你可以参加一至两个培训班,跟着教练以确保运动得当,这会有很大的好处。要记住,即使不去健身房,你也可以在家里或公园里进行许多免费活动。上网查一下,会有成千上万种免费的创新运动项目。关键是要能持之以恒,并随着时间的推移增加强度或运动时长,你就会体验到体能的改善。

参加群体锻炼是结识新朋友的好方法,包括徒步旅行和骑行俱乐部、武术学校、滑冰队、舞蹈团,等等。如果你想结交新朋友,参加体育活动是结识各色人等的好方法。如果你觉得和别人一起锻炼更容易,那就加入一个群体。你手机上有一些应用程序可以帮忙找到这样的集会,或者也可以上网搜索。如果看到一群人在活动,不要害羞,直接上前请求加入。如果你最终不喜欢这一集会,可以权且持续一段时间,直到找到契合自己的群体为止。也许你需要加入

CHAPTER 9
Exercise

for the benefits until you find the right match. Maybe you will have to join six to eight groups before finding the one that best fits your personality.

When I was young, I was an exercise fanatic. I would take every opportunity to break a sweat. I carried a gym bag and was always ready for a work out. As I got successful in business, I started making excuses for not exercising.

I am not sure what got me started again, but one day I got back into exercising after a long break. I realized something. I found that when I worked out, my consumption discipline improved greatly. I am not sure why this happens, but I think it is because doing one good thing for your body may cause the brain to have a spillover effect to other areas. Exercising helps you stay focused, especially on the diet. I have mentioned this to many people, and most agree that physical activity provides enhanced mental discipline for staying on a diet. For this reason, a regular exercise program works hand in hand with *The Baby Food Diet*.

Getting started on a daily exercise routine is the crucial first step. I advise working out at least three days a week and no more than five. The body needs rest to benefit from exertion. I always advise changing up your routine every six weeks. This does not mean a complete change, but instead you should modify a few things in your program. For example, if you are swimming, then change

第九章
运动

6~8个群体才能找到最适合自身个性的那一个。

我年轻的时候,是一个运动狂。我会抓住每一个机会大汗淋漓。我会随身携带一个健身包,随时随地准备去健身。而随着事业的成功,我开始找各种借口不去锻炼。

我不确定是什么原因,只是有一天,在长时间的中断之后,我又重新开始健身了。我意识到一些东西,发现当我锻炼的时候,我的饮食自律有了很大的提高。我不确定为什么会这样,但我认为这是因为对自己身体做一件好事会导致大脑对其他区域产生溢出效应,锻炼能帮助你保持专注,尤其是在节食方面。我向许多人提到过这一点,而且大多数人都认同,体育活动可以增强坚持节食的心理自律。因此,经常锻炼身体与**婴儿食品节食法**可以相辅相成、共同作用。

开启日常锻炼的习惯是至关重要的第一步。我建议每周至少锻炼3天,但不要超过5天,身体需要休息才能从运动中受益。我总是建议每6个星期改变一下日常习惯,这并不意味着要全盘改变,但是应当修改健身计划中的一些内容。例如,如果你在游泳,那就改变泳姿;如果你正在骑自行车,也许可以尝试不同于

the type of stroke you are doing. If you are riding a bike, maybe try a different route than in the past. If you lift weights, make sure you add intensity over time to keep getting stronger.

My career is in investment banking. On several occasions while exercising I was able to come up with solutions to problems. What happens during physical activity is that your mind frees itself from the usual thought cycles and a mental clarity is achieved. On multiple occasions, intense training has indirectly made me lots of money. Out of nowhere and for no reason, like magic, a solution pops into my head while exhaustively sweating, and the next thing I know, the deal gets done. During exercising, I am no longer consciously thinking about the business parameters or the road block to closing it. Then suddenly a resolution comes to mind. When I go back to work, I know what to do and proceed toward closing that transaction.

I know many people who have had similar experiences while physically engaged. Besides putting the mind at rest and allowing you a chance to relax, exercising reduces stress for sure. Most doctors will tell you that stress is one of the main reasons people have health problems in the first place. So, for me, working out is so much more than building muscles or accelerating my heart rate; it gives my brain a much-needed break from the grinds of my normal everyday schedule. There is no justifiable reason not to at least add walking to your

第九章
运动

过去的路线；如果你正练习举重，一定要随着时间的推移增加强度，让自己变得更强壮。

我的职业是投资银行从业者。有好几次在健身的时候，我想出了解决工作中问题的办法。体育活动会让大脑从常规的思维循环中解放出来，从而获得清晰的意识。很多时候，高强度的体能训练间接地让我赚了很多钱。不知从何而来，没有任何理由，就像魔术一样，在我汗流浃背之时，一个解决方案突然出现在我的脑海中，然后我所知道的下一件事，便是交易达成了。在健身的过程中，我不再有意识地去思考具体交易，然后，一个解决方案突然凭空出现了。当我回去工作时，我就知道该做什么，并着手完成那笔交易。

我知道很多人在从事体力活动时也有类似的经历。除了让你的大脑得到休息，有机会放松，毫无疑问，锻炼还能减小压力。很多医生会告诉你，压力是人们健康问题的首要原因之一。所以，对我来说，锻炼不仅仅能增肌或促进心率，还让我的大脑从每天细碎的日程安排中得到了急需的休息。没有任何正当的理由不把散步添加到日常活动中。如果你和我一样野心勃勃，那就去健身房吧！

daily activities. If you are more ambitious like me, join a gym.

Meditation can be an effective addition to your health routine. Many people find meditation also relieves stress and helps obtain clarity in their thoughts. I do not practice formal meditation as outlined by many books, but I do benefit greatly from this practice several nights each week while listening to music or silently lying in bed. I let my mind rest, focus on my breathing, and enjoy the music or stillness. I do this when I am completely alone.

If you want to combine exercise and meditation, I understand that some types of yoga offer both procedures. I have not studied yoga in depth, but I have taken a few classes just to experience the essence of it and have observed many people benefiting. Whether you are talking about exercise or meditation, there is no one way or style that is best. There are hundreds of books and classes readily available. Find something you like and just go with it. Maybe mix it up: tennis on Mondays, the gym Tuesdays, and hiking on Fridays. Maybe yoga twice a week complemented by swimming three times. Just experiment with what works best for you.

When I lived in Los Angeles, the traffic was horrendous. Driving somewhere that should take 15 minutes could take 1.5 hours. I needed to be at my office by 8 AM. To get there on time I needed to leave my home at 6:40 AM. However, if I left at 5:30 AM, it only took

第九章
运动

冥想可以是健康习惯的有效补充。许多人发现,冥想还能减轻压力,并助其理清思绪。我没有按照很多书中描述的那样正式地练习冥想,但我确实从每周几个晚上听音乐或静静地躺在床上的冥想中受益匪浅。让大脑休息,专注呼吸,享受音乐或宁静。当我独自一人的时候就会这么做。

如果要把运动和冥想结合起来,我知道有些类型的瑜伽同时提供这两种形式。我没有深入学习瑜伽,但是上过几节课,只是为了体验瑜伽的精髓,并注意到很多人从中受益。无论说到健身还是瑜伽,没有一种方式或形式是最好的,有数百种书籍和课程可供选择。找自己喜欢的,然后直接参加。也许可以混合一下:周一打网球,周二去健身房,周五去远足。也许一周做两次瑜伽,再配上三次游泳。尝试最适合自己的方式。

我住在洛杉矶的时候,那时交通状况非常糟糕。开车去一个 15 分钟就能到达的地方,可能需要 1.5 个小时。我需要在早上 8 点以前到办公室,要想准时到达,就得在早上 6:40 出家门。然而,如果我早上 5 点半出门的话,只需要 15 分钟就能到。我的交通补救措施是

CHAPTER 9
Exercise

15 minutes. My traffic remedy was to join a gym near my office. I joined one that was open 24 hours a day. Every day, I left my house between 5 AM and 5:30 AM. The night before, I would put my work clothes in the car. In the morning, I would wear my gym clothes and head straight to work out. After exercising and enjoying my new, like-minded friends, I would shower and shave there. The most important benefit was that every day started off in a positive way. There was no getting frustrated by fighting morning traffic, and after the workout, I felt great every single day. This was a creative solution to a bad situation.

Who wants to sit in traffic? Not me for sure. I like to start my day off with success. If you follow a routine like this, you can pre-blend a protein shake and bring the baby food along to the gym. After the workout, you enjoy them both and set off to work. What a wonderful way to start a day.

Make sure as part of the diet, and especially if you are exercising, to drink plenty of water. A good solution to stifling hunger is to drink one or two glasses of water before you eat your baby food. This will make you feel full and satisfied after your small meal. When starting the diet, as I mentioned, the first two weeks are the most challenging. This is true with all diets because your body takes time to unlearn old habits and readjust to new ones. That's why it's crucial to simply drink a full glass of water

第九章
运动

在办公室附近找一个健身房,我找到了一家24小时营业的。我每天早上5点到5点半出门,前一天晚上,我会把工作服放进车里。第二天早上,我穿上运动服,直接去健身。在锻炼身体并享受与志同道合的新朋友在一起之后,我会在那里洗澡和修面,最重要的好处是每天都是以积极的方式开始的。不再因为早晨堵车而感到沮丧,锻炼后,我每天都感觉很棒。这是对于糟糕情况的创造性解决方法。

谁愿意堵在路上?反正我不愿意。我喜欢以成功开始新的一天。如果你也照这样做,可以预先搅拌一份蛋白奶昔,和婴儿食品一起带到健身房。锻炼之后,就可以享受这两种食物,然后开始工作。这是多好的开始一天的方法啊。

要确保作为节食的一部分,一定要喝足够多的水,特别是如果你一直在健身。抑制饥饿的一个好办法就是在吃婴儿食品之前喝1~2杯水,这会使你在吃完自己的小份餐食之后有饱腹感并很满足。正如我之前提到的,开始节食的前两周是最具挑战性的。所有的节食方法都是如此,因为你的身体需要时间来忘掉旧习惯,重新适应新习惯。这就是为什么在每餐之

or two before each meal to fight natural hunger pangs. After two weeks on *The Baby Food Diet*, your body will begin to adjust and everything will be much easier. However, the practice of drinking a glass or two of water before eating is something that works before and after your body adjusts. Especially once you are on the maintenance part of the diet, you may want to follow this approach with all your meals.

Many diets involve joining support groups with likeminded people on the same regimen. This is not a requirement to succeed, but it is indeed a good idea. If you have others on the same diet and experiencing the same cravings, it is much easier to stick with the diet. You can share ideas and thoughts and help each other to realize success.

In today's world, you need not be in the same town or country for this to work. Form a group that chats on your smartphone or is connected via an app. Share your success story with others. People like to share success. That is one of the reasons we like to have friends. It is human nature to want to interact with others. To support and help someone else is rewarding for both people involved. Think about how good it feels to be recognized and complimented for your achievements. Think about how good it feels to help others and to add value to another person's life. Start a group or join an existing group with the same goals to foster a supportive environment.

第九章
运动

前喝 1~2 杯水来对抗自然的饥饿感,是至关重要的。在坚持**婴儿食品节食法**两周后,身体会开始适应,一切都会变得更容易。饭前喝 1~2 杯水的习惯是在身体调整之前和之后都有效的,特别是在节食的维持期阶段,你可能每顿饭都想遵循这一方法。

许多节食方法中都会提到加入与志同道合者组成的互助小组。这不是成功的必要条件,但确实是一个好主意。如果有其他人和你一起节食,并且有同样的期望,那就更容易坚持下去。你们可以分享主意和想法,帮助彼此实现成功。

在当今世界,人们不需要在同一个城镇或国家来使群体发挥作用。从智能手机的应用程序上组一个聊天群,与他人分享你的成功故事。人们喜欢分享成功,这是我们喜欢交朋友的原因之一。想要与他人互动是人类的天性,支持和帮助他人对双方都是有益的。想想你的成就得到认可和赞美的感觉有多好,想想帮助他人并为他人的生活增添价值的感觉有多好。建一个群或加入一个有相同目标的群,以营造一个可支持的环境。

CHAPTER 10

Losing Weight Requires Commitment

第十章

减肥需要承诺

CHAPTER 10
Losing Weight Requires Commitment

One of the best features of *The Baby Food Diet* is convenience. As mentioned in the previous chapter, I travel with my baby food, protein, and a portable blender. However, you can be less hard core than me. Just about every city in the world has stores that sell baby food. When traveling, you can substitute the protein shake with a can of tuna fish which, like baby food, can be purchased just about anywhere you are visiting. One of the keys to success with *The Baby Food Diet* is to be flexible in your approach.

It is important not to give yourself any excuses for failure. You must decide that no matter what, you will stick with the diet in one form or another. If your travels are so hectic that you fall off the diet for a few days, that is certainly not the end of the world. Just start again as soon as possible. Do not delay or be upset with yourself for temporarily ignoring the diet as this will happen occasionally. After the break, just forgive yourself and begin again.

These days, everyone talks about science. People use scientific theories to create ideas and hype about your health and to ultimately sell their products. Look at the gym supplements craze: protein, pre-workout, post workout, breakfast, and sleeping enhancements. Each one has an associated study, each one has "proof". Let me tell you something, there are a lot of perfectly healthy people who don't take any

第十章
减肥需要承诺

婴儿食品节食法的特色之一就是方便。正如前一章所提到的,我旅行时随身携带婴儿食品、蛋白粉和一个便携式搅拌机。不过,你不必像我这么固执、教条。世界上几乎每个城市都有售卖婴儿食品的商店。当你旅行的时候,可以用一罐金枪鱼来代替蛋白奶昔。金枪鱼和婴儿食品一样,在你去的任何地方都可以买到。**婴儿食品节食法**能够成功的关键之一就是方式灵活。

重要的是不要给自己任何失败的借口。你必须下定决心,无论如何,都要坚持这种或那种形式的节食。如果旅行特别繁忙,以至于中断节食几天,那也肯定不是世界末日,只要尽快重新开始。不要由于自己暂时忽视了节食而继续拖延或烦恼,这种情况偶尔会发生。在中断之后,要原谅自己,并重新开始。

如今,人人都在谈论科学。人们用科学理论来创造关于健康的想法和进行宣传,并最终销售其产品。看看健身营养补充剂热潮:蛋白质营养补充剂,锻炼前营养补充剂,锻炼后营养补充剂,早餐和增强睡眠营养补充剂。每种都经过相关的研究,每种都有其依据。让我告诉你事实,有很多非常健康的人从不吃任何营

supplements. The problem is that everyone is looking for the fastest and easiest approach.

A pill that helps you drop weight is the goal of everyone in the pharmaceutical industry. The easiest road to success is what drives people to consider options like liposuction. I think this is not a healthy approach and somewhat over the top. I also believe there is something wrong with the current approach of a wonder pill for losing weight. It is obvious that *The Baby Food Diet* is the easiest and most prudent way to manage your weight, mostly because you will never be hungry.

What will it take to motivate you to get started today? What do you need to do to convince yourself to begin right now? Here are some practical and persuasive answers. Take off your shirt and look in the mirror. Do you like what you see? Look at your old high school pictures when you were thin; that should be good motivation. Think of the girl or guy you are attracted to, but they never noticed you. Think about the low energy levels you feel, and why you avoid doing things that you would really like to do. How about your self-confidence? Do you like the way your clothes fit? The list goes on, so go ahead and take a moment and try these things. Get on the horse and get started on the ride to success. Be free of self-doubt and soar to new heights. That's what *The Baby Food Diet* will do for you. Be your own master and stop

第十章
减肥需要承诺

养补充剂。问题在于,每个人都在寻找最迅速、最简单的方法。

一种有助于减肥的药丸是每个制药企业的生产目标。最简单的成功方法就是促使人们考虑像抽脂这样的选择。我认为这不是一种健康的方法,而且有点太过头了;我还认为,目前那种使用减肥特效药的方法也有问题。很明显,**婴儿食品节食法**是控制体重最简单和最明智的方法,主要是因为你永远不会感觉到饿。

怎样才能激励你从今天开始呢?你需要做些什么来说服自己从现在就行动呢?以下是一些实用且有说服力的答案。脱下衬衫,照照镜子,喜欢你所看到的吗?看看老照片中高中时苗条的自己,这应该是很好的动机;想想你喜欢的女孩或男孩,而他们却从来没有注意到你;想想感觉到身体的低能量水平,以及为什么不去做真正想做的事情;你的自信心如何?你喜欢所穿衣服的松紧度吗?这样的例子不胜枚举。所以,不妨花点时间直接去尝试一下。骑上马,并开始奔向成功之路。摆脱自我怀疑,飞向新的高度。这就是**婴儿食品节食法**对你的作用。做自己的主人,停止拖延。

procrastinating.

How many times have I heard the excuse that someone is big boned, so they can never be thin? Stop making excuses! The fact is, I am big boned. Sure, people have different body types. However, if you are fat, you are fat. Even big boned people will look like stealth warriors after they undertake *The Baby Food Diet*. No excuses anymore, begin now. Later you will laugh at yourself for not starting sooner.

第十章
减肥需要承诺

我曾听过无数次这样的借口：声称某人骨架大，所以他们永远不会变瘦。别再找借口！事实上，我也是大骨架之人。当然，各人有各自不同的体形。然而，如果你胖，就是胖，即使是骨架大的人，在尝试**婴儿食品节食法**后，也会像隐形战士一样。别再找借口了，现在就开始！之后你会自嘲没能早点儿开始。

CHAPTER 11

Body, Life and Balance

第十一章

身体、生活和平衡

CHAPTER 11
Body, Life and Balance

I have lived in China on and off for many years. As a direct result, I learned that the body needs to be in balance. We already talked about meditation and exercise, but what about acupuncture? Acupuncture can be effective in getting your body in balance as you drop the pounds. It can also help in your mental attitude as it rebalances your physical elements.

I am not sure how or why acupuncture works, but I have observed it working wonders for some of my friends. It seems that when the acupuncture needles are placed in the skin at specific points on the body, endorphins are released. I am told that these endorphins have a calming and relaxing influence on your mind and body. The result is that it is easier to deal with stress, frustration and anxiety. We all know that binge eating can come as a direct result of stress. I am told that the acupuncture treatments will improve your digestion and increase your metabolism which enhance weight loss. Some people even think acupuncture can increase your willpower.

I have had acupuncture many times. First, you will be told that it does not hurt and is not painful. I disagree, as each time I have had acupuncture I have experienced pain. For this procedure to be effective you must properly communicate what exactly is bothering you with your acupuncturist. Next, they will check your pulse to get a reading of your energy. Then they look at the condition

第十一章
身体、生活和平衡

我断断续续在中国生活了很多年,直接结果就是明白了身体需要保持平衡。我们已经讨论过冥想和健身,那么针灸呢?在减肥时,针灸可以有效地使身体保持平衡。由于它会重新平衡身体元素,因此也可以帮助改善精神状态。

我不确定针灸是如何发挥作用以及为何能发挥作用的,但曾观察到它为我的一些朋友创造了奇迹。似乎当针灸的针尖扎入身体特定部位的皮肤里时,内啡肽就会释放出来。我听说这些内啡肽会对身心有一种平静和放松的作用,其结果是更容易处理压力、挫折和焦虑。我们都知道暴饮暴食可能是压力造成的直接结果。我听说针灸疗法可以改善消化系统和加速新陈代谢,从而促进减肥,有些人甚至认为针灸可以增强意志力。

我扎过很多次针灸。首先,你会被告知一点都不疼。我不同意,因为每次扎针灸的时候我都经历了疼痛。为了使针灸这一过程更有效,你必须正确地与针灸师沟通,到底是什么毛病在困扰你。接下来,他们将检查你的脉搏,以获得能量数据。然后他们会观察你舌头的状况,以获取线索,决定你应接受什么样的治

of your tongue which gives them clues as to what treatment you are to receive. I am no expert and only speaking from my personal experience, so I am not sure of exactly what they are looking for, but this is the process.

At this point, the treatment will start. Let me repeat, it is a little painful. In fact, for me, it is very painful. The acupuncturist picks areas based on the ancient knowledge of "Traditional Chinese Medicine". One acupuncturist also asked me to start deep breathing to stimulate the results. That makes sense as it is a well-known fact that concentrated breathing is a way to control pain. For example, during child birth a mother will focus on intense breathing techniques to ease pain.

During these visits, I also learned about pressure points that can be stimulated to promote better health. I am told there are multiple pressure points in our bodies. One is just behind each ear and is said to relieve stress. Another is inside the elbow where the skin creases when your arm is bent and is practiced to reduce anger. It is easy to find, because when pressure is applied here the pain will be extreme. A third painful pressure point is between your thumb and index finger and is said to relieve tension and pain in your back and neck. There are many more pressure points which you can learn about if research this very interesting subject.

One day, I was talking with a friend, who is a

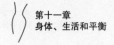

第十一章
身体、生活和平衡

疗。我不是专家,只是从我个人体验来说,我不确定他们到底在寻找什么,但就是这样一个过程。

此时,治疗就开始了。我再说一遍,真的有点疼。事实上,对我来说,非常疼痛。针灸师根据古老的中医知识来选择针灸区域。一位针灸师还让我开始深呼吸以促进效果。这是有道理的,因为众所周知,深度集中呼吸是一种控制疼痛的方法。例如,在生小孩的时候,母亲会专注于高强度呼吸以减轻疼痛。

在几次扎针灸的过程中我还了解到,可以通过刺激穴位促进健康。我听说身体有很多个穴位,有一个就在耳朵后面,据说可以缓解压力;还有一个在肘部内侧,当手臂弯曲时,那里的皮肤会起皱,按压这个穴位可以减少怒气,而且这个穴位很容易找到,因为当你用力按压时,那里会特别疼痛;第三个减缓疼痛的穴位在拇指和食指之间,据说按压可以缓解背部和颈部的紧张和疼痛。如果去研究这个非常有趣的课题,可以了解更多的穴位。

有一天,我正和一个中医朋友聊天,突然我开始咳

Traditional Chinese Medical doctor, and I suddenly started coughing in an uncontrollable manner. He grabbed my hand and applied pressure. Immediately, I stopped coughing.

If this is interesting to you, I advise either visiting a qualified acupuncture specialist and/or buy a detailed book on the topic. Acupuncture treatment and especially pressure points are fascinating to me as I know that they work in treating people.

As noted, I have experimented with acupuncture and, while it is not a replacement for dieting, there is no harm in complementing your diet with some acupuncture treatments. While I say there is no harm in trying acupuncture, realize that you may experience a little pain.

I believe acupuncture helped me in better balancing my body and will benefit you as well. However, realize it is a complement, just like exercise, that enhances *The Baby Food Diet*, not a replacement. Any such heath related procedures combined with *The Baby Food Diet* can help you achieve your goals.

As mentioned earlier in the book, I often take twelve plus hour flights. The biggest problem with such flights is not the loss of time. Most long-haul airplanes now have Internet connectivity. Nowadays, I can complete business chores during my flight. I also use the time to meditate and relax.

第十一章
身体、生活和平衡

嗽,是无法控制的那种。他抓住我的手给我按压穴位,我立刻停止了咳嗽。

如果你对此感兴趣,我建议你要么去拜访一位有资质的针灸专家,要么买一本详细的相关书籍。针灸治疗,尤其是按压穴位,令我非常着迷,因为我知道真的有疗效。

如前所述,我曾经扎过针灸,虽然它不能代替节食,但用针灸疗法作为节食的补充并没有害处。虽然我说扎针灸没有害处,但要知道可能会感到一点疼痛。

我相信针灸能助我更好地平衡身体,同样也会对你有好处。然而,要意识到这只是一种补充,就像健身一样,可以加强**婴儿食品节食法**的效果,但不是替代品。任何一种与健康相关的方法,结合**婴儿食品节食法**,都可以帮助你实现目标。

正如书前面所提到的,我经常乘坐12个多小时的飞机。飞这种航班最大的问题不是时间的浪费,现在大多数长途飞机都有互联网连接。如今,我可以在飞行中完成商务琐事,也可以利用这段时间进行冥想和放松。

The problem I encounter with such long flights is jet lag. Adjusting sleeping patterns can help with overcoming jet lag. However, changing my sleeping habits is far from enough to solve this problem. I am told that on such long flights the atmospheric conditions cause many health issues, such as bloating and muscle cramps. These things in turn cause a person's body to be less efficient, lethargic and even bring on more serious illness.

My friend from England, David Thorne introduced me to a solution to overcome jet lag that has worked wonders for me. I now rarely experience jet lag thanks to David's teachings. In fact, what I learned and experienced is that this technique is also beneficial when dieting.

David is married to a wonderful Chinese woman named Ying. During their 8+years in China, Ying introduced David to Traditional Chinese Medicine ("TCM"). One technique David learned from his exposure to TCM is something I will refer to as "Slap Yourself to Good Health". David plans on publishing a book on this subject in the future. I look forward to reading the book once it is available.

For now, let me give you a glimpse of some basics of his regimen. The slap routine is based around about a dozen very simple actions that work from the head down to release the muscles and invigorate the systems. It doesn't just work on the tangible muscles and blood flow

第十一章
身体、生活和平衡

我在长途飞行中遇到的问题是时差。调整睡眠模式可能有助于克服时差反应。然而,改变睡眠习惯远远不足以解决这个问题。据听说,在这样的长途飞行中,机舱内通风条件会导致许多健康问题,比如身体肿胀和肌肉痉挛。这些问题反过来会导致身体效率降低、昏昏欲睡,甚至导致更严重的疾病。

我的英国朋友大卫·索恩(David Thorne)向我推荐了一种克服时差反应的方法,对我产生了神奇的效果。多亏了大卫的教导,我现在很少有时差反应。事实上,我学到和体会到的是,这种方法在节食时也是有益的。

大卫娶了一个很棒的中国女人,名叫英(Ying 音译)。在他们八年多的中国生活中,英向大卫推荐了中国传统医学"TCM"。这是大卫在接触传统中医的时候学到的一个技巧,我将其称为"拍打经络,有益健康"。大卫计划将来出版一本关于这一主题的书,我期待着此书的问世,希望去阅读。

在目前情况下,让我向你们简单介绍一下这种方法的一些基本情况。拍打动作基于十几个非常简单的动作,这些动作从头部向下进行,可以放松肌肉,激活系统。拍

but also dips into what we in the west see as the intangible and perhaps mystical energy systems. I am referring to the Meridians, which also leads us onto the Yin and Yang concepts.

David watched and listened and by default over the years put together these 12 actions taken from Doctors, Monks and various healers throughout the Provinces in China. You can see the elderly in some of the village squares doing some of the regimen.

He picked up one action from one person and then noticed a slightly different approach from another person in a completely different part of China and then began to establish a pattern. So perhaps it was from his unique position as a traveler from the west that he was able to figure this out.

All you do is vigorously slap and stimulate different parts of your body for about 30 seconds to one minute in each established area. I suggest waiting until you are in a quiet place as it can be quite noisy. It starts with some simple stretches, which release and inspire flow of blood and energy. Then it simply follows on from the head region and works down attending to various and specific parts of the body (arms, legs, etc.).

David suggests that this very ancient regimen stimulates not only blood flow but also generates energy through the meridians, which are now being taken more seriously by the western medical authorities.

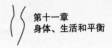

第十一章
身体、生活和平衡

打不仅作用于有形的肌肉和血液流动，还渗透到我们西方人认为的无形的，或是神秘的能量系统中，我指的是经络，它也能引导我们进入阴阳的概念。

多年来，大卫一直在观察、倾听，并默默地将中国各省的医生、僧侣和各种治疗师采取的12种动作汇总在一起。你经常可以看到一些老年人在村里的广场上做一些这样的动作。

大卫从一个人那里学到了一个动作，然后他注意到，在中国完全不同地区的另一个人，也会用略微不同的方法，然后他开始创建一种拍打方案。所以，也许正是出于他作为一个来自西方的旅行者的独特角度，才能够想得出来。

你所要做的就是用力拍打和刺激身体的不同部位，在每个部位持续拍打30秒到1分钟。我建议你要到一个安静的地方，因为声音可能会很大。先从一些简单的伸展运动开始，释放和激发血液流动和能量。然后简单地从头部区域开始拍打，向下延伸到身体的各个特定部位（手臂、腿等）。

大卫认为，这种非常古老的疗法不仅能刺激血液流

The results are more than amazing. As mentioned, I now rarely experience jet lag. Even better, I have added the slapping as part of my daily workout. At the end of each session I take five minutes and go through the slap routine. According to David this is like magic and makes the workouts much more effective.

Look for David's book on this slap routine in the future.

I had a former girlfriend who had the best idea ever. She had found a doctor who would help lose weight through massage therapy. I love getting a massage, so we went ahead and tried this strategy. Of course, there are many benefits that come from a vigorous massage: relaxation, stress reduction, detox, and improved blood circulation.

However, massage therapy did nothing to help me lose weight. It was so much fun that I would gladly do it again regardless of the results. The massage therapist would give me a rubdown and would proceed to knead my stomach in circles for fifteen minutes. After that, I entered a sauna, sweated profusely, and shed lots of water weight. Obviously, water loss does not count as real fat loss. In the end, it was enjoyable, and I loved the experience despite not losing any real fat.

第十一章
身体、生活和平衡

动,还能通过经络产生能量,而现在西方权威医学也开始对经络更加重视起来。

效果特别惊人。如前所述,我现在很少有时差反应。更好的是,我把拍打经络作为日常锻炼的一部分。每节课结束时,我都要花5分钟进行拍打。据大卫说,这就像魔术一样,会使健身更有效。

将来,你可以在大卫的书中找到关于拍打经络的方法。

我的前女友曾有一个特别好的主意,她找到了一位可以通过按摩疗法帮助减肥的医生。我喜欢做按摩,所以我们就一起去尝试了这个方法。当然,强力按摩有很多好处:放松、减压、排毒和改善血液循环。

然而,按摩疗法并没有帮助我减肥。按摩非常有意思,不管结果如何,我都愿意再去。按摩师会给我做全身按摩,然后在我的腹部做15分钟的转圈按摩。之后,我去蒸桑拿,大汗淋漓,减少了很多水分。很明显,失水并不算真正意义上的减脂,但最后还是令人愉快的,我喜欢这种经历,尽管没有减掉任何真正的脂肪。

CHAPTER 12

Alcohol and Supplements

第十二章

酒精和营养补充剂

CHAPTER 12
Alcohol and Supplements

There is an old saying: "Time is going to pass anyway, so why not do something now?" If you have long considered doing something about your weight, now is the time to begin. You can try all the TV commercially-hyped or newfound fad dieting programs, but why not just try the permanent solution? Why not start *The Baby Food Diet* today before time passes by you?

Envision how fantastic it will feel to wake up each day, take a shower, and look at the skinny image in the mirror? I have a winning technique — a type of reward program that can help with your journey of losing weight. Is there something you have wanted to buy for a long time? Possibly you can afford the item, but doing so might inhibit your finances. You just need a sensible reason to justify the purchase. Why not make it a gift to yourself for success? Write down a promise that you will buy the gift if you stay on *The Baby Food Diet* for 30 days. The truth is, the money you save may well be enough to afford what you wanted anyway. I mentioned it before and will say it again: *The Baby Food Diet* will bring down your food costs dramatically, since baby food is not expensive and certainly much cheaper than going out for dinner.

As the diet progresses, I advise scheduling subsequent rewards for yourself as you reach key dates. I say dates, not weight goals. The amount of weight loss will vary from person to person. Different factors determine how

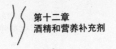

第十二章
酒精和营养补充剂

俗话说:"时间一去不复返,做事为何不趁早?"如果你长期以来一直在考虑减肥,现在是时候开始了。你可以尝试所有电视上炒作的广告或新流行的节食节目,但为什么不尝试永久的解决方案呢?为什么不从今天就开始**婴儿食品节食法**呢?

想象一下,每天醒来洗个澡,看着镜子里纤瘦的自己,这种感觉该有多棒?我有一个成功的技巧,是一种奖励计划,可以助力你的减肥之旅。有没有什么东西是你一直想买的?也许你买得起,但买之后可能会有些财务吃紧,只是需要一个值得购买的合理理由。为什么不把它作为成功的礼物呢?写下一个承诺,如果你能坚持**婴儿食品节食法**30天,就去买这个礼物。事实上,单是你省下来的钱很可能就足够买想要的东西了。我之前提到过,还会再说一遍:**婴儿食品节食法**将大大降低你的食物成本,因为婴儿食品并不贵,肯定要比出去吃饭便宜得多。

随着节食的进展,我建议在达到关键期限时为自己安排后续奖励。我说的是期限,不是减肥目标。所减的重量会因人而异,不同的因素决定了减肥的速度,这并不像消耗了卡路里那么简单。对一些人来说,坚持90

CHAPTER 12
Alcohol and Supplements

quickly you will lose weight. It is not as simple as the calories you consume. For some people, sticking with the Baby Food Reduction Diet for 90 days is enough, but for others, it may take 180 days or even more. What matters is that you will reach your target goals with perseverance and staying with the diet. How long it takes is not the definitive objective; beginning and diligently continuing with the process is all that matters.

If you choose the Baby Food Fast, you should follow a few rules. You can eat two baby foods and one protein shake per day. This must be spread over three meals, not eaten all at once. It does not matter what meal you eat first as long as during the day you consume only two baby foods and one protein shake. Also, you must drink lots of water. The Baby Food Fast is for those in a hurry and will allow you to lose weight rapidly. This diet should never last more than 21 days (14 days is best). For health reasons, after a maximum of 21 days you will need to a break and switch to the Baby Food Reduction Diet. The Baby Food Fast is an exceptional boost for quick weight loss.

As you read this book, you have undoubtedly noticed that I keep on repeating the importance of getting started, so why not just do it today? I do have a reason for repeating myself. I figure if I remind you every few pages, then the encouragement will prompt you to commit and begin.

第十二章
酒精和营养补充剂

天的婴儿食品减食期就足够了;但对另一些人来说,可能需要180天,甚至更长。重要的是,通过坚持不懈与努力节食,你会达到目标体重。需要多长时间并不是最终的目标,重要的是对于节食能够着手开始并持之以恒。

如果选择婴儿食品禁食期,应该遵循一些规则。你可以每天吃两份婴儿食品和一份蛋白奶昔。这些必须分散到三餐中,而不是一下都吃完。只要你每天只吃两份婴儿食品和一份蛋白奶昔,先吃什么并不重要。此外,你必须饮用很多水。婴儿食品禁食期是为那些急需瘦身的人准备的,能让你快速减肥。这一节食阶段不能超过21天(最好是14天)。出于健康原因,最多21天之后,需要中止,改为婴儿食品减食期。婴儿食品禁食期是能够快速减肥的特殊促进法。

当你看这本书的时候,肯定已经注意到我一直在重复"着手开始"的重要性,所以为什么不从今天就开始呢?我确实有理由重复自己的话。我想,如果每隔几页就提醒一次,那么这种鼓励就会促使你承诺并开始。

CHAPTER 12
Alcohol and Supplements

Be careful with substituting while on *The Baby Food Diet*. For example, apple sauce is not baby food. It has lots of sugar and may cause you to gain weight. Stick with basic baby foods. Often other foods that are in a puree form may have significant amounts of added sugar and are not suitable for *The Baby Food Diet*. Also be cautious regarding smoothies from restaurants which are often loaded with sugar and high in calories. It is not unusual for a juice shop to add unhealthful ingredients to enhance the smoothie's taste. Ask what is in the drink before ordering.

People inquire about drinking alcohol while on the diet. For me, the worst part of imbibing alcohol is that it stimulates my appetite. I can be cruising along fine and not cheating on the diet, but if I start drinking, I tend to start making poor food choices. For whatever reason, when I consume alcohol, I desire the foods I should not be eating. I often listen to the old cravings and give in to the temptation. To enjoy alcohol from time to time is not wrong, but moderation should be the barometer.

There are many experts who say drinking two glasses of red wine each day is healthy. This may be true, but I do not really know. To me, it sounds like a solid excuse for promoting the drinking of red wine daily. By my definition, if you are only drinking two glasses of wine each day, that does not require an excuse, because drinking in moderation is okay.

第十二章
酒精和营养补充剂

在**婴儿食品节食法**中,要小心替代品。例如,苹果酱不是婴儿食品,它含有大量的糖分,可能会导致体重增加。要谨遵基本的婴儿食品。通常其他以泥状形式存在的食物,可能含有大量的添加糖,不适合**婴儿食品节食法**。另外,也要小心餐馆里的冰沙,因为它们一般含有大量的糖分和很高的热量。对于果汁店来说,添加不健康的成分来增加冰沙的口味是很正常的,点餐前问一下饮料里有什么。

有人询问在节食时饮酒的情况。对我来说,喝酒最糟糕的地方就是它会刺激食欲。在节食中,我可以一直很好地遵循,不会作弊,但如果我开始喝酒,就会开始做出糟糕的食物选择。不管出于什么原因,当我喝酒时,就想吃那些不应该吃的食物。我总是听从旧的渴望,并屈服于诱惑。偶尔饮酒并没有错,但适度饮酒应该是晴雨表。

有很多专家说,每天喝两杯红酒是有益健康的。这可能是对的,但我真的不清楚。对我来说,这听起来像是一个促使每天喝红酒的好借口。根据我的定义,如果你每天只喝两杯酒,并不需要借口,因为适量饮酒是可以的。

CHAPTER 12
Alcohol and Supplements

The biggest problem I have with drinking alcohol is that I tend to lose productivity the next day. My energy level is reduced, and I am not nearly as productive. If I drink too much, I am sure that my focus and decision-making ability is hindered. I am adamant about not drinking and driving and putting your life in danger along with the innocent lives of others. Always have a designated driver when consuming alcohol.

Many times, I have heard the term: "Be a responsible drinker." I am not sure what this means. We drink to have fun, relax, and forget about the pressures we face, not to take on more responsibility. The idea of relaxing is probably the best argument for occasional drinking as it puts you in a different place for the time being. However, there are other ways to achieve the same result. Try meditation, a walk on the beach or long kiss with a lover.

Like I said, I am not against liquor, in fact I drink in moderation myself. My advice is uncompromising about drinking when on the Baby Food Reduction Diet or the Baby Food Fast. There is no room for alcohol! However, when you are on the Baby Food Maintenance Diet, it is okay to drink moderately.

Everyone in the world knows the negative effects of drinking too much alcohol. It is not healthy. However, many things in life are not healthy. *The Baby Food Diet* is healthy, and if you stay on it, you can occasionally do

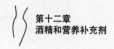
第十二章
酒精和营养补充剂

我喝酒的最大问题在于第二天工作效率会下降。我的精力水平降低了，工作效率也基本没有。如果我喝得太多，我的注意力和决策能力肯定会受到阻碍。我绝不酒后开车，不让自己的生命和其他人的无辜生命处于危险之中。酒后一定要找一个司机代驾。

我听过很多次这样的话："做一个负责任的饮酒者。"我不清楚其确切含意。我们喝酒是为了娱乐、放松，和忘记所面临的压力，而不是承担更多的责任。放松的想法可能是偶尔喝酒最好的理由，因为它会让你暂时置身事外。然而，还有其他方法可以达到同样的效果。可以试试冥想，在沙滩上散步或与爱人长吻。

就像我说的，我并不反对喝酒，事实上我自己也适度饮酒。我的建议是绝不能在婴儿食品减食期或婴儿食品禁食期内饮酒。要禁酒！然而，当你处在婴儿食品维持期时，适量饮酒是可以的。

世界上每个人都知道过量饮酒的负面影响，是不健康的。然而，生活中的许多事情都是不健康的。**婴儿食品节食法**是健康的，如果你一直坚持下去，可以偶尔做一些不健康的事情。这是因坚持**婴儿食品节食法**而获得

unhealthy things. You have earned this right by being on *The Baby Food Diet*. Having a drink now and then or even over indulging once a month does not stop you from enjoying the benefits of *The Baby Food Diet*.

Enough on alcohol, unless you think you are an alcoholic. If so, please seek help. There are many people and organizations who will assist you if you take the first step and reach out. If not properly treated, alcohol can destroy your life.

I hear much about superfoods like Apple Cider Vinegar which is made by crushing apples and squeezing out the liquid. Yeast is added to start the fermentation process that creates alcohol. In a second process, the alcohol is turned in to vinegar. The taste is very sour, and the smell is not good.

Some believe that Apple Cider Vinegar has no nutritional value. I disagree. There are several health benefits to adding Apple Cider Vinegar to *The Baby Food Diet*. It is said Apple Cider Vinegar can aid with weight loss, regulate blood sugar levels, lower cholesterol, improve skin health, and reduce symptoms of acid reflex. The list of claims goes on and on. It is my opinion that Apple Cider Vinegar cannot hurt you, and it has helped me to achieve my goals.

Never drink Apple Cider Vinegar straight, always dilute it. One tablespoon three times a day I am told, is

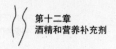
第十二章
酒精和营养补充剂

的权利。偶尔喝一杯,甚至每月放纵一次,都不能阻止你享受**婴儿食品节食法**的好处。

酒的问题就不多说了,除非你认为自己是个酒鬼。如果是这样,请寻求帮助。只要你迈出第一步并主动求助,会有很多人和组织帮助你。如果处理不当,酒精会毁了你的生活。

我听说过很多像苹果醋这样的超级食物,是通过碾碎苹果并挤出汁液制成的。加入酵母,开始发酵过程,并产生酒精;经过二次发酵,酒精变成了醋。味道奇酸,而且难闻。

有些人认为苹果醋没有营养价值。我不同意。在**婴儿食品节食法**中加入苹果醋有几个健康益处。据说苹果醋有助于减肥,调节血糖水平,降低胆固醇,改善皮肤健康,减少胃酸倒流状况。此清单可以无限扩展。我认为苹果醋不会有害,而且帮助我实现了目标。

千万不要直接饮用苹果醋,一定要稀释。据说一天三次,每次一汤匙,是最佳剂量。然而,我也听说,每天

the optimum amount. However, I have also heard that just one tablespoon per day will be sufficient to produce positive results. To increase the benefits to your digestion and microbiome, always use raw unfiltered organic apple cider vinegar.

Another supplement you may want to add to your daily diet is aspirin. The idea of taking one low dosage aspirin each day to avoid heart attacks and strokes seems to be based on intense scientific studies. The cost of aspirin is next to nothing and could be the one pill that extends your life. I am not a big believer in taking too many drugs but an aspirin can have a potentially positive impact on overall cardiovascular health.

This brings me to genetics and knowing your family history regarding health. It is a fact that DNA plays a huge part in your future well-being. If you are from a family where relatives have historically died at an early age due to illness, you should pay attention. It is possible that you may be a candidate for similar health problems. If you are such a person, please start *The Baby Food Diet* now. You owe it to yourself and to your family to live a long and healthy life. *The Baby Food Diet* is a decisive first step to reaching old age.

Even if all previous members of your family lived to be one hundred years old, that does not make you invincible. Be thankful and appreciative for robust bloodlines, but it is still prudent to take additional steps to

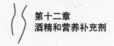

只要一汤匙就足以产生积极的效果。为了增强消化系统和有益菌群,应该经常服用未经过滤的原生有机苹果醋。

另一种你可能想要添加到日常饮食中的营养补充剂是阿司匹林。每天服用低剂量阿司匹林以避免心脏病和中风的想法似乎是基于大量的科学研究。阿司匹林的价格很低,而且可能是延长寿命的一种药物。我不太推崇服用过多的药物,但阿司匹林可能对心血管整体健康有潜在的积极影响。

这让我想到遗传学以及想去了解家族健康史。事实上,DNA在你未来的幸福生活中扮演着重要的角色。如果你的家族历史上曾有亲人因疾病而早逝,就应该多加注意,你也很有可能出现类似的健康问题。如果你有这种情况,请从现在就开始尝试**婴儿食品节食法**。你有责任为自己和家人过上健康长寿的生活。**婴儿食品节食法**是走向长寿决定性的第一步。

即使你的家族以前所有的成员都活到100岁,这也不能使你所向无敌。要对长寿的血统心存感恩与感激,但

ensure a long and healthy future. Why not add the insurance of *The Baby Food Diet* to that blessed heredity group, "The Lucky Gene Club", and enjoy an even healthier lifestyle? We can all feel better, be more productive, and have a happier life as a direct result of following *The Baby Food Diet*.

What about the effect of sun on your skin and overall welfare? The sun is not as beneficial as we once thought. When I was young, I was taught that a tan was the mark of a healthy individual. We now know differently. Actually, the sun causes skin damage that cannot be repaired. In China, people are well aware of this truth, and in the summer, you see people using umbrellas to avoid the harmful sun rays. So, please take the time to shield your skin. Exposure to the sun is good in small dosages and can help you achieve a positive mental attitude. Sunshine makes people feel happy and has been credited with several health benefits, including the easing of depression. While the sun does not directly give off vitamins, it begins the process that allows your body to make its own vitamin D. Wear a hat and spread on sunscreen to protect your skin and to stay youthful and beautiful.

Speaking of vitamins, how beneficial are they? On my first job after college, I was determined to be a success by working as smart and hard as possible to achieve my goals. I always got to the office before anyone else and

第十二章
酒精和营养补充剂

采取额外的措施来确保长久和健康的未来仍然是明智的。为什么不把**婴儿食品节食法**这一保险添加到"幸运基因俱乐部"这一幸福的遗传群体中,享受更健康的生活方式呢?作为遵循**婴儿食品节食法**的直接结果,我们都可以感觉更好,更有效率,会有更快乐的人生。

那么阳光对皮肤和整体健康会有什么影响呢?太阳并不像我们曾经认为的那样有益健康。我年轻的时候,被灌输皮肤晒成古铜色才是人体健康的标志。我们现在有了不同的认识。事实上,太阳对皮肤造成的伤害是巨大的。在中国,人们很清楚这个道理,一到夏天,你会看到人们用遮阳伞来避免有害的紫外线。所以,请花点时间保护皮肤。少量的阳光照射是有益的,可以帮助你获得积极的心态。阳光使人感到快乐,并公认对健康有几大好处,包括减轻抑郁。虽然阳光不会直接产生维生素,但经过晒太阳这一过程,能让身体自行生成维生素 D。戴上帽子,涂上防晒霜,保护肌肤以永葆青春和靓丽。

说到维生素,能有多大益处呢?在我大学毕业后的第一份工作中,我下定决心要通过尽可能聪明和努力的工

was the last to leave. My extra effort was rewarded as I was soon earning more money than anyone I knew. However, I noticed at times I was feeling a little tired. After all, I was usually in the gym by 5:30 AM, at my office by 7:30 AM, and often worked till late at night. As a precautionary measure, I made it a point to eat well-balanced, wholesome meals and went to my doctor for regular checkups. He knew how hard I worked and realized the daily stress I was under and informed me that even though I was eating in a healthy way, my body was not getting enough of the vitamins that are needed to combat enormous stress. He reasoned that vitamins would keep me healthier and give me more endurance.

He suggested I start taking a multivitamin and gave me a sample. He cautioned that it takes about six weeks for vitamins to start to have an impact. The positive results of vitamins are so subtle that most people do not even notice the difference. If a person stops taking vitamins, after a few weeks they often experience less energy and wonder why. Vitamins are not a quick fix to better health, but I believe that they do have long-term benefits.

I recommend taking a high-quality multivitamin every day, whether you notice the difference or not. Scientific research suggests organic compounds slow aging and keep our bones strong. We all know that calcium strengthens skeletal mass, but calcium cannot do this alone.

第十二章
酒精和营养补充剂

作来实现目标,获得成功。我总是比别人先到办公室,而且是最后一个离开的。我的加倍努力得到了回报,因为我很快就赚到了比认识的任何人都多的钱。然而我注意到,有时会感到有点累。毕竟,我通常早上 5:30 到健身房,7:30 到办公室,而且经常工作到深夜。作为预防措施,我坚持吃均衡健康的食物,并定期去看医生,做常规体检。医生知道我工作有多努力,也意识到我每天承受的压力。他告诉我,尽管我吃得很健康,但我的身体并没有获得足够的维生素来对抗巨大的压力。他阐明维生素会让我更健康,让我更有耐力。

医生建议我开始服用复合维生素,并给了我一个剂量。他提醒我,维生素大约需要六周的时间才能开始发挥作用。维生素的积极作用是如此微妙,以至于大多数人甚至根本发觉不到差别。如果一个人停止服用维生素,几周后,他们通常会感到精力不足,并奇怪为什么。维生素不能快速改善健康状况,但我相信它确实有长期的益处。

不管你是否感觉到其中的差别,我建议每天服用高质量的复合维生素。科学研究表明,有机化合物可以延缓衰老,保持骨骼强健。我们都知道,钙可以增强骨骼

Nutrients such as calcium, vitamin K, vitamin D, and magnesium work together to fortify our calcified framework. This is a great argument for taking a daily multivitamin which works in unison with a variety of complementary vitamins.

There is evidence that vitamins can reduce the chance of heart disease by releasing antioxidants that shield your heart from inflammation. Vitamins can also moderate other aspects of your being. For example, Vitamin D plays a role in creating happiness. Vitamin B Complex is excellent for reducing stress. Vitamin A, Vitamin C and Vitamin E are great for your skin. Vitamin B is said to give you healthier hair. One of the prime benefits of taking vitamins on a regular basis is that they boost your immune system, making you less prone to getting sick. An example of this is Zinc which is said to combat colds and other infections. Some people believe that a multivitamin also reduces depression caused by lack of sleep, weakness and other symptoms of lethargy.

Vitamins are called supplements because they complement your diet. They are not meant to replace nutritious meals. *The Baby Food Diet* will provide the healthy ingredients your body requires and, with the addition of a multivitamin, you should have all the nutrients needed. With all the positive assertions of vitamins and nothing to lose, please add a multivitamin to your daily regimen and never stop.Best of all, even the top

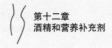
第十二章
酒精和营养补充剂

质量，但钙不能单独做到这一点。钙、维生素K、维生素D和镁等营养素共同作用，才能防止骨骼钙化。这是每天服用复合维生素的最好理由。

有证据表明，维生素可以通过释放抗氧化剂来降低罹患心脏病的概率，而抗氧化剂可以保护心脏免受炎症的侵害。维生素还可以调节身体的其他方面。例如，维生素D在生成幸福感中发挥作用；复合维生素B对减轻压力非常有效；维生素A、维生素C和维生素E对皮肤有好处；据说维生素B能使头发更健康。定期服用维生素的主要好处之一是，能够增强免疫系统，使你不容易生病。锌就是一个例子，据说锌可以对抗感冒和其他感染。一些人认为，复合维生素还能减缓因睡眠不足、虚弱和其他嗜睡症状引起的抑郁。

维生素被称为营养补充剂，因为它们可以作为饮食的补充，但它们并不是要取代营养丰富的膳食。**婴儿食品节食法**将提供身体需要的健康成分，再加上复合维生素，你就能得到所有需要的营养素。所有的维生素都是积极有益的，没有负面影响，请在日常饮食中添加复合维生素，一直不要停止。最好的是，即使是高端的复合维生素也是可以负担得起的。要记住，

multivitamin is affordable. Remember, when striving to be in good shape, every little thing you do can make a big difference in creating better overall health and leading to a longer, happier life.

第十二章
酒精和营养补充剂

当你努力保持好身材的时候,你所做的每一件小事都能对整体健康产生巨大的影响,让你拥有更长久、更快乐的人生。

CHAPTER 13

100 Amazing Suggestions

第十三章

100 个绝佳的建议

CHAPTER 13
100 Amazing Suggestions

By now, I hope you are convinced to embrace *The Baby Food Diet* revolution -- your forever diet. The resulting achievements are numerous. Physically, you will be thin and in better physical shape. Mentally, you will be more self-confident and exude a new outlook on life that will improve your social skills and appreciation of yourself and others.

Additionally, your friends and family will stop worrying about you. *The Baby Food Diet* will benefit all aspects of your life.

Here are my 100 suggestions that can make you a healthier person:

1. Count your blessings every day, and take the time to realize that you are special and unique. Tell yourself that you make a positive difference in the world. Truly believe that you can achieve all your goals.

2. Appreciate the good things in your life, including the smaller, less often celebrated aspects that contribute to your well-being.

3. Balance your life. Having balance has been one of my life's greatest challenges. I tend to focus too much on one thing and lose equilibrium because I have not slowed down enough to recognize the problem. To maintain symmetry, go for a walk once a week or try some form of mindfulness to help rebalance your priorities.

第十三章
100个绝佳的建议

事到如今，我希望你已经被说服了，去尝试**婴儿食品节食法**这一革命方法——它是你永远的节食方法。由此产生的成果数不胜数。身体上，你会更瘦更健康；精神上，你会更加自信，焕发新的人生活力，将会提高你的社交能力并欣赏自己和他人。

此外，你的朋友和家人会停止为你担心。**婴儿食品节食法**将有益于生活的方方面面。

以下是我的100条建议，可以让你成为一个更健康的人：

1. 每天都要怀有感恩之心，花时间认识到自己是独一无二的。告诉自己，你在这个世界中做出了积极的改变。真正相信自己能实现所有的目标。

2. 欣赏生活中的美好事物，包括那些对你的幸福贡献得较小的、不值一提的方面。

3. 平衡你的生活。保持平衡是我人生中的挑战之一。我很容易由于过于关注一件事而失去平衡，因为我没有放慢脚步来认识这一问题。为了保持平衡，可以每周

4. Always improve yourself. Commit to becoming a better person instead of remaining stagnant. You can begin by learning something new or starting an exercise program. There are instructional classes, books, blogs, online materials, and so many other avenues to help expand your base of knowledge which will all result in self-improvement. For now, the most important task is starting *The Baby Food Diet*.

5. Learn to listen more instead of constantly talking. It is hard to learn anything when you are always yapping away. The key is to slow down which will in turn force you to take a deep breath and clearly hear every word spoken to you. After you have absorbed the message, take a moment to decipher its meaning before responding. Take notes and be open to new ideas that are different from your current way of thinking.

6. Be kind to others. Take the time to genuinely care about people around you. This does not take much time or effort. Sometimes a warm smile or friendly hello will go a long way in making somebody else's day.

7. Respect leaders and other people who have power. I frequently notice that people say negative things about leaders. This is not a good approach to success. Instead, it is better to listen and learn by being open to other opinions. You do not have to agree with everything,

第十三章
100个绝佳的建议

散步一次,或者尝试用一些正念的形式来帮助重新平衡你的优先级。

4. 要不断提升自己。致力于做一个更好的人,而不是停滞不前。你可以以学习新东西或参加健身作为开始,培训班、书籍、博客、在线教材,以及许多其他的途径可以帮助你扩展知识基础,这些都将有助于自我完善。目前来讲,最重要的任务是开始**婴儿食品节食法**。

5. 学会多听少说。你要是一直喋喋不休,很难学到任何东西。关键是要慢下来,这反过来又会迫使你深呼吸,清晰地听明白别人对你说的每一个字。在你接收了他人表达的信息之后,在回答之前,花点时间去理解其真正含意。要做笔记,接受与你目前思维方式不同的新想法。

6. 善待他人。花点时间真正关心你周围的人,这并不需要花费太多的时间和精力。有时候,一个温暖的微笑或友好的问候会让别人开心一整天。

7. 尊重领导和其他当权者。我经常注意到人们说领导的坏话。这不是一个成功的好方法。相反,更好的方法

but it is an easier and smarter road to success if you stay within the box as opposed to being a rebel.

8. Expect the best of yourself. That means being punctual, well-groomed, and dressing in a becoming way that shows respect for whomever you are meeting. Live for success which is exhibited by the way you handle yourself in front of others.

9. Never assume success, but instead make the extra effort to achieve that goal. Think about what can go wrong before it happens and have a solution already in place to deal with all potential challenges.

10. If you start an exercise program, learn to do the movements correctly so as not to hurt yourself. Warmup by stretching so your muscles are loose and prepared for whatever activity you are undertaking. Also, make sure you have the proper shoes to avoid injury.

11. Get a good night's sleep, so you will wake up with a winning attitude. This is an ideal time for self-talk; believe me it works. When you awaken and before you retire are perfect times to read aloud your list of affirmations.

第十三章
100个绝佳的建议

是倾听并接收他人观点,从中学习。你不必事事赞同,但如果你能循规蹈矩,而不是当反叛者,这会是一条更容易、更聪明的成功之路。

8. 期待最好的自己。这就意味着要守时,打扮整洁,穿着得体,无论你要见谁,这都是尊重对方的体现。为成功而活,这表现在你在他人面前把握自己的方式。

9. 永远不要假设成功,而是要加倍努力去实现这一目标。事前要未雨绸缪地想想可能会出现什么问题,并且预先做好解决方案,以应对所有潜在的挑战。

10. 如果你开始一项锻炼计划,就要学会正确地做动作,以免自己受伤。通过伸展运动来热身,使肌肉放松,为即将进行的所有活动做好准备。此外,确保你有合适的鞋子,以避免受伤。

11. 睡个好觉,这样醒来的时候你就会有一种必胜的姿态。这是理想的自言自语时间,相信我,真的管用。在醒来后和入睡前,是大声朗读自我肯定内容的完美时间。

12. If you do something wrong, whatever it is, just forgive yourself. It does not matter if it is something as simple as cheating on the diet or something bigger, you must forgive yourself and move forward.

13. Stay away from vindictive people. If you sense that a person is malicious, you should immediately cut them out of your life. Such people will eventually distort something important in your life and cause you major problems, so instantly distance yourself.

14. Take the time to put your family first. They rely on and love you. Return that love and respect by continually showing how much you care about them. This does not mean you have to agree with them, or do as they say, you just need to appreciate them.

15. Always remain calm in all situations. Never lose your cool or show anger as it never creates success. Being openly angry only leads to failure. If you lose self-control, you lose the respect of others. There may be a good reason for your anger, but a temper tantrum will never solve the problem.

16. Take time out for yourself. Success begins and is sustained by staying in touch with who you are.

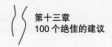

12. 如果你做错了事,不管是什么,要原谅自己。不管是简单的节食期间作弊还是更大的事情,你都必须原谅自己,继续前行。

13. 要远离有报复心之人。如果你感觉到一个人心怀恶意,应该立即把他从你的生活中剔除出去。这样的人最终会扭曲生活中一些重要的东西,给你带来大麻烦,所以你要马上远离。

14. 花些时间把家人放在第一位。他们依赖你并且爱你,通过不断地表达你对他们的关心来回报他们的爱和尊重。这并不意味着你必须同意他们的观点,或者照他们说的去做,你只需要对他们心存感恩。

15. 在任何情况下都要保持冷静。永远不要发火或愤怒,因为这样不会使你成功,公然愤怒只会导致失败。如果你失去自我控制,就失去了别人的尊重。你可能有很好的生气理由,但是发脾气永远解决不了问题。

16. 给自己留点时间。成功始于且持续于了解真实的自己。

17. Add garlic to your diet in the winter. Garlic has certain antiviral properties that can boost your immune system.

18. Echinacea can help you with overcoming a cold and will also strengthen your immune system. I carry it with me when traveling.

19. I used to think small talk was a waste of time. I now realize that forging straight into business matters can only lead to failure. I suggest enjoying what appears to be meaningless chatting because such nonchalant communication reveals personalities, character, and integrity that can create lasting friendships.

20. I said it many times, and here I go again: drink at least 8 glasses of water every day.

21. Use a calendar to stay organized and punctual. Do not count on remembering everything. I have a tendency to enjoy most people in meetings and often lose track of time and my next appointment. Therefore, before entering a meeting, I set an alarm clock to assure I am not late to the next one.

22. Cayenne pepper will improve your circulation. When you are on *The Baby Food Diet*, it is okay to add small amounts of cayenne pepper to your shake if it doesn't adversely affect you.

第十三章
100个绝佳的建议

17. 在冬天的饮食中加入大蒜。大蒜有一定的抗病毒特性,可以增强免疫系统。

18. 紫锥菊能够有助于战胜感冒,还能增强免疫系统。我旅行时会随身携带。

19. 我过去认为与人闲聊是浪费时间。现在我意识到,直入正题只会导致失败。我建议享受看似毫无意义的聊天,因为这种漫不经心的交流,能够显现出可以建立长久友谊的性格和为人。

20. 我说过很多次了,现在还得说:每天至少要喝8杯水。

21. 使用日程表来保持条理性和守时性。不要指望单凭脑子记住一切。我乐于在开会时享受和多数人在一起的感觉,经常忘记时间和之后的约会。因此,在参加会议之前,我会设置一个闹钟来确保自己不会耽误下面的行程。

22. 辣椒能促进血液循环。在**婴儿食品节食法**期间,如果没有什么副作用的话,可以添加少量辣椒到搅拌物中。

23. If you have a headache, rub your temples with peppermint oil. If you suffer headaches often, you can conveniently carry peppermint oil in your bag.

24. Listen to music to relax. Music has magical healing qualities that can put you in a positive mood. Contemporary music will help you stay in touch with the times. Whether you like it or not, force yourself to listen to 10 minutes each day. You will learn much about current affairs which will keep you updated and educated. That does not mean you should stop enjoying oldies or classics, but it is a wise to stay in touch with today's crowd.

25. If you want something, develop a strategy to obtain it. Maybe that requires learning new techniques, so an education will be in order. Possibly it means more practice time to master a new skill. Whatever it entails to be successful, you can make it reality with a proper strategy and the right mind set.

26. Never give up too easily. Often on my road to success, I have failed. However, I always immediately examine why I failed. Maybe it was just bad timing, and I can wait it out and try again later. Maybe I was talking to the wrong person and should try again with someone else. The point being, never take the first failure to be an indication that your objective is unworthy. Often, there is another path that will lead to success. Sure, there are times to quit and move on, but do not give up too soon.

第十三章
100个绝佳的建议

23. 如果感觉头痛，可以用薄荷油按摩太阳穴。如果你经常头痛，可以把薄荷油放入包里，这样更方便。

24. 倾听音乐放松自己。音乐具有神奇的疗愈作用，能让你心情愉悦。当代音乐有助于你与时俱进。不管喜欢与否，强迫自己每天听10分钟，你会学到很多新鲜事物，这将使你不断更新自己并学到知识。这并不意味着你应该放弃享受老歌或经典歌曲，但时刻与当今外界保持联系是明智之举。

25. 如果你想要什么，制定一个策略以期获得。也许需要学习新技术，所以教育是必要的。这可能意味着需要很多的时间去练习才能掌握一项新技能。无论成功需要什么，你都可以通过适当的策略和正确的心态使之成为现实。

26. 永远不要轻易放弃。在通往成功的路上，我常常失败。然而，我总是立即审视失败的原因。也许只是时机不对，我可以过些时候再试；也许我找的人不对，应该再换其他人试试。重要的是，永远不要把第一次失败看成是该目标不值得实现的标志。通常，总会有

My experience dictates trying at least one more time before totally conceding on any idea.

27. Checklists are a wonderful tool to stay better organized and super-efficient. Forgetting things and being unorganized are stressful. Before I used checklists, I would waste time and money buying things that were not necessary. Now that I follow checklists, I never forget anything which has relieved stress and the frustration created when I overlooked significant issues.

28. Designate specific places to store specific things when you are done with them. I used to lose my cell phone a lot. One day, I had enough of this scatterbrained behavior and started putting my phone in the same pocket after every use. I got in the habit of touching that pocket each time I changed locations to assure I had not forgotten the phone. I have not lost a phone in ten years.

29. Find a few good friends who will be there when you feel weak. It seems even the most casual of conversations can help you relax and forget about the stress you are experiencing. Friends can help you stay positive and feel better about yourself.

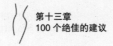

第十三章
100个绝佳的建议

另一条路可以通往成功。当然，有时候你可以放弃，然后继续前行，但不要过早放弃。根据我的经验，在完全放弃任何想法之前，至少要再试一次。

27. 清单是一个极好的工具，可以让你保持更有条理和高效。忘记事情和没有条理会令人倍感压力。在使用清单之前，我会浪费时间和金钱购买不必要的东西。现在我按照清单行事，再也不会忘记任何事情，也因此减轻了压力，以及因忽视重要问题所产生的挫折感。

28. 在使用完某件物品之后，要把它们放到特定的地方。我以前经常丢失手机。有一天，我受够了这种粗心大意的行为，开始决定每次用完手机后都把它放在同一个口袋里。我养成了这样的习惯，每当离开某一地方时，都要摸一下那个口袋，以确保我没有忘记拿手机。我已经有十年没再丢过手机了。

29. 找几个好朋友。当你感到虚弱无助的时候，他们会待在你身边的那种。似乎即使是最随意的谈话也能帮助你放松，忘记所承受的压力。朋友可以帮助你保持积极向上，让你自我感觉更好。

30. Always be conscious of doing little things to improve your health, like walking up the stairs instead of riding the elevator or escalator. As another example, consider lifting dumbbells or pedaling the stationary bike when watching television.

31. When encountering difficult situations and challenges, slow down and first identify the positive aspects of the circumstances. Then move forward with a solution.

32. Take small breaks throughout the day no matter how busy you are. If you keep working with no pauses, you will lose perspective and, ultimately, your efficiency will suffer as well. For me, taking a break means slowing down and relaxing my mind and body which often prompts a novel solution to ongoing complexities. We call that an "ah-ha" moment.

33. Join meet-up groups in your spare time to experience different activities and hear varying opinions from newfound friendships.

34. Take the time to notice when someone is doing an outstanding job. For example, if a hotel employee goes above and beyond to accommodate me, I will take a few moments to write a positive comment on the establishment's social media or send management a short note of appreciation. It is true that what goes around does come around.

第十三章
100个绝佳的建议

30. 要经常有意识地做一些小事来改善自身健康,比如爬楼梯而不是乘电梯或自动扶梯。另一个例子是,当你看电视的时候,可以考虑举哑铃或者脚踏动感单车。

31. 当遇到困境和挑战时,要放慢脚步,首先确定环境的积极方面,然后找到解决方案继续前行。

32. 不管有多忙,每天都要适当小憩一下。如果你一直不停地工作,就会失去洞察力,最终,你的效率也会受到影响。对我来说,休息意味着放慢脚步、放松身心,这样做往往能为正在进行的复杂工作提供一个新奇的解决方案,我们称之为"灵感"时刻。

33. 在业余时间里加入聚会小组,体验不同的活动,从新结识的友谊中听取不同的意见。

34. 花些时间去注意他人出色的工作。例如,如果酒店员工的优质服务超出了我的预期,我就会花一些时间在该酒店的社交媒体上给出好评,或者给管理层发一封简短的感谢信。的确,好人会有好报。

35. In addition to a checklist, write a "To-do" list. It will keep you more organized and best of all it will allow you to prioritize.

36. Laugh at yourself, but never at others. It is disrespectful to laugh at other people and can cause you to lose friends and business opportunities. To laugh at yourself is a win in every way possible. You will learn not to be embarrassed which is a useless emotion that can deter you from following up on something important. By laughing at yourself, you will also learn not to overreact to situations. To overreact is to lose.

37. Never live in the past; the past is gone. However, if you are feeling down, it can be therapeutic remembering something successful that you accomplished previously. It can help re-ignite the same disposition you need to succeed again.

38. Never be a "know-it-all" as that is a quick road to failure. No matter how much you know, it is never a good strategy to win a meaningless argument or to prove someone is wrong and embarrass them, especially in the company of others.

39. Switch up your daily routine to stay fresh. It is funny, but as early as grade school, kids take a seat in a classroom and never change to another chair. If you go to the same lunch place every day, why not try sitting at a different table? There are so many minor changes that can lead you to be more open to success. Routine is good, but it

35. 除了清单外,写一个待办事项表,会让你更有条理。最重要的是,能让你分清轻重缓急。

36. 可以自嘲,但绝不要嘲笑别人。嘲笑他人是不礼貌的,会让你失去朋友和商业机会。在任何可能的情况下,自嘲都是一种胜利,你将学会不再尴尬。尴尬是一种无用的情绪,会阻止你去做一些重要的事情。通过自嘲,你也能学会不要对一些状况反应过度。反应过度就是失败。

37. 永远不要活在过去,过去的已经过去了。然而,如果你感到沮丧,可以回想一下自己以前达到的成就,这是有疗愈作用的,可以帮助你重新点燃再次成功所需要的同样的情绪。

38. 千万不要做一个"万事通",因为这是通往失败的捷径。不管你知道多少,想要赢得一场毫无意义的争论,或者证明某人是错的并让他们难堪,这从来都不是一个好策略,尤其是在别人面前。

39. 改变你的日常生活,保持新鲜。早前上小学的时候,孩子们坐在教室里,从来不换座位,这样很好笑。如果你每天都去同一个地方吃午餐,为什么不试着换到另

can also stifle other opportunities and cause you to run through life with your eyes closed.

40. Take the time to applaud others, rather than seeking to expose their faults. I see jealous people criticize successful people all the time. Instead, try to appreciate and emulate their achievements by realizing that you, too, can reach phenomenal accomplishments in the future.

41. Remove negative people from your life. If you know someone who is constantly bringing you down, it is wise to avoid such pessimism. Life is challenging enough without someone regularly pointing out your flaws.

42. This sensory tip works for me, but not for everybody. I like things to smell nice and often use scents to brighten up my attitude. When I enter a building that is pleasantly aromatic, I immediately comment and automatically feel better. When I encounter a bad odor, I only want to leave. Just as with good music, scents can have an enchanting or unpleasant influence on your mood.

43. Passion is paramount in life. If you can add passion to your daily activities, you will notice that everything in your world is positively impacted.

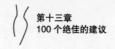

一张桌子旁边坐呢？有很多微小的改变可以使你更有可能成功。例行公事是好事，但它也会扼杀其他机会，让你闭着眼睛过完一生。

40. 花些时间为他人喝彩，而不是试图揭露他们的错误。我看到嫉妒的人总是批评成功人士。相反，应该通过意识到你也能在未来取得非凡的成就，而试着去欣赏和模仿他们的成就。

41. 让负能量的人远离你的生活。如果你认识一个人，总是让你情绪低落，明智的做法是避免这种悲观情绪。生活本身就够有挑战性的了，不需要别人经常指出你的缺点。

42. 这个感官小窍门对我很有用，但不是对每个人都有用。我喜欢闻好闻的东西，会经常用香水来提升状态。当我进入一座芳香怡人的建筑时，我会立即发表评论，并自然而然地感觉很好。当我闻到难闻的气味时，我只想离开。就像好听的音乐一样，气味会对你的情绪产生迷人或不爽的影响。

43. 激情在生活中至关重要。如果你能在日常活动中

44. If you find yourself thinking negative thoughts, slow down and figure out what is causing the pessimism. Retrain your mind to think in a more positive and productive manner. When I was young, I had a wicked temper. I never got upset about big things, but the little ones drove me crazy. It was embarrassing. I knew I needed to gain control of my temper, so when I became angry, I began to study myself and analyze the reasons it was happening. For example, if I was standing in a food line and had had a bad experience in the past, I would let it trigger my thoughts. I would think, if this happens again, I will be very angry. Of course, when it happened, I got very angry. Remember earlier when I said you are what you think? When I at last realized my problem, it was so easy to control my temper. Now in such a situation, I will either order something different or maybe just be polite and not get angry, if the same situation happens again. The moral of the story is you must control your thoughts. Think long and hard to determine the root of this angry behavior. What thoughts are you putting in your mind that are the catalyst to creating the anger? Be patient as it took me about 8 months to realize what triggered my temper. Be persistent, and you will eventually uncover the origin of the negative thoughts and find a solution.

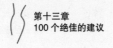

第十三章
100个绝佳的建议

加入激情,就会发现你的世界里的一切都受到了积极的影响。

44. 如果你发现自己在想消极的事情,那么放慢脚步,找出导致悲观情绪的原因,重新训练大脑,使其以一种更积极有效的方式思考。在我年轻的时候,有一个坏毛病,从不为大事烦恼,但小事会把我逼疯。这很尴尬。我知道需要控制自己的脾气,所以当我生气的时候,我开始研究自己,分析事情发生的原因。举个例子,如果我在排队买食物时,曾有过不好的经历,这就会触发我的联想。我会想,如果这种事情再次发生,我将非常生气。当然,当事情真的发生时,我确实非常生气。还记得我之前说过其思即其人吗?当我终于意识到自己的问题时,就很容易控制自己的脾气了。如果同样的情况再次发生,我要么点不同的食物,要么礼貌客气,但绝不生气。这个故事的意思就是在于你必须控制自己的思想。好好想想,找出这种愤怒情绪的根源。你头脑中有哪些想法是引发愤怒的催化剂?要有耐心,因为我花了大约8个月的时间才意识到是什么触发了我的脾气。要坚持不懈,你最终会发现消极思想的根源,并找到解决办法。

45. It is imperative to be resilient in life as bouncing back from failure is an integral part of reaching success. When you fail, you must jump up and try again. On the road to riches, there will always be a myriad of stumbling blocks that must be dealt with along the way. It is absolutely essential to persist despite what people around you are advising. Realize that each failure brings you closer to your ultimate goals as long as you remain unshakeable and optimistic. Success can be easy, but is normally a constant challenge with many obstacles along the way. Resilience and persistence must be part of your DNA to achieve your goals and dreams.

46. Learn to be a savvy salesperson. Sales is nothing more than persuading someone to buy something. No matter what you do in life, it involves selling yourself or a product through any number of learnable techniques. Often, you must convince people to see your line of thought by providing analysis. Combine that with a little encouragement and a friendly attitude to help influence another person. Something as small as persuading friends to go where you want is being a salesperson. Procuring a date with the person of your dreams is being a salesperson. Your presentation will determine whether or not you succeed. Learn sales and soar to new heights in business as well as in your personal life. Sometimes the greatest sell you will make is to convince yourself to do

第十三章
100个绝佳的建议

45. 在生活中一定要具有柔韧性,因为从失败中恢复回来是成功不可或缺的一部分。当你失败时,一定要振作起来,再试一次。在通往财富的道路上,沿途总会有无数的绊脚石需要你去跨越。不管周围的人怎么建议,坚持自我是绝对必要的。要意识到,只要你保持坚定和乐观,每次失败都会让你离最终目标更近。成功可能会很容易,但通常需要不断的挑战,路上还会有许多障碍。为了实现目标和梦想,柔韧性和持久性必须成为你DNA的一部分。

46. 要学会做一个精明的推销员。推销无非是说服别人买东西。无论你在生活中做哪一行,都需要通过技巧来推销自己或产品。通常,必须通过提供种种分析来说服人们看到你的思路,再加上一点鼓励和友好的态度来帮助你影响他人。像说服朋友去你想去的地方这样小的事情就是在做推销员;实现和梦中情人约会也是在做推销员。你的演说将决定自己是否成功。学习营销,在生意场上和个人生活中达到新的高度。有时候,说服自己去做某件事是最大的推销。我的意思是停止拖延,现在就开始节食。

something. What I mean by this is stop procrastinating and start the diet now.

47. Develop a reliable and trustworthy team of like-minded individuals. In the beginning you may have to wear many hats, but as time goes on a team can be helpful. While you should understand all aspects of your life and business, that does not mean performing all tasks yourself. Delegating effectively and overseeing your mission is a smart way to maximize your efforts. You only have 24 hours in a day, so trust your team to make deadlines and enjoy the ensuing success.

48. Never think you are so important that you are above doing any job. Conversely, be as knowledgeable as possible in regards to every task related to your job which gives you the latitude to pick up the slack when needed. This familiarity serves multiple purposes, including earning the respect of those around you. Let's take two large restaurants. In one, the owner never helps with anything, like bussing a table, believing he is above it all. From what I have observed, an operation with management like that will soon go out of business. Now let's look at another large eatery where the owner sees people waiting for a table. He also sees that all the servers are busy, so this owner immediately cleans the table himself and seats the waiting customers. This type of hands on operation will normally flourish and reach

47. 培养一个由志趣相投之人组成的可靠和值得信赖的团队。刚开始创业的时候，你可能要身兼数职，但随着时间的推移，团队合作会很有帮助。虽然你应该了解自己在生活中和事业上的方方面面，但这并不意味着你要自己完成所有的任务。将任务有效地委派给他人并进行监督是使你的努力最大化的聪明方法。你一天只有 24 小时，所以要相信你的团队能在最后期限前完成任务，并享受随之而来的成功。

48. 永远不要认为自己特别重要，不屑于做任何工作。相反，对于每一项与工作相关的任务，你都要尽可能地博学多才，这样在需要的时候，你才有足够的能力去填补空缺。这种业务上的精通有多种用途，包括赢得周围人的尊重。让我们以两家大饭店为例。其中一家，老板从不帮忙做任何事情，比如从不帮助清理餐桌，他认为自己凌驾于一切之上。据我所知，像这种管理运营方式，餐厅很快就会倒闭。现在我们来看看另一家大饭店，当老板看到食客在等位，并看到所有的服务员都很忙时，这位老板马上亲自清理了餐桌，并让等候的食客入座。这种类型的运营通常会使企业蓬勃发展，并达到成功的新高度。原因很简单：由于老板无私的职业道德得到

new heights of success. The reason is simple: the employees and customers learn to respect the owner due to his or her selfless work ethic. Employees will always go the extra mile for someone personally involved like this, and consumers will frequently return to support such an impassioned owner.

49. Find a good mentor, who has traveled the path you are currently following, to help navigate your destination. If you want complete success, seek out several mentors in the different areas of your life. These counselors can lend advice and encouragement. Maybe consult one mentor regarding health, another for exercise, a third for wealth management, etc. These battle-tested guides can help you avoid pitfalls in addition to introducing you to influential people who can expedite your ascension to unlimited heights.

50. To be successful, you must have a passion for your vision and radiate it wherever you go. Passion is a true belief and is contagious. If you really believe in something, others around you will believe and feel excited about being part of the team. Passion also provides that extra bit of energy to endure whatever it takes to realize success. A passionate person exudes confidence, and confidence creates victory.

51. Write notes instantly when ideas pop up. I used to have a problem forgetting details of imagined concepts. This started when I was a child. At one point, a wise friend

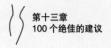

了员工和客户的尊重。员工们总是会为这种亲力亲为的人而加倍努力,而消费者也会经常回来支持这样一位充满激情的老板。

49. 找一位曾走过你现在所走之路的导师,帮助你找到目标。如果你想获得完全的成功,那就分别在生活中不同领域内找几位导师。这些导师可以提供建议和鼓励。也许你会向一位导师咨询健康方面的问题,向另一位咨询健身方面的问题,向第三位咨询理财方面的问题等等。这些经过实战考验的导师们可以帮助你避免陷阱,还能把你介绍给有影响力的人,使你加速提升到无限的高度。

50. 要想成功,你必须对自己的愿景充满激情,无论走到哪里,都要将激情散发出来。激情是一种真实的信念,而且非常具有感染力。如果你真的相信某件事,你团队中的人就会相信,并为成为其中的一员而感到兴奋。激情也提供了额外的能量去忍耐为实现成功所需要的一切。一个充满激情的人会散发出自信,而自信创造胜利。

51. 当想法出现时,立即写下来。我曾经出现过这样

told me to keep a pad and pencil at my bedside for the times when I awaken after an enlightening dream. Now I use my phone to record any new ideas when I am out and about. At the end of each week, I review my notes and recordings and never forget that next billion-dollar idea or strategy.

52. The "Eye of the Beholder" can be a confusing platitude in business and personal life. During meetings, I have experienced hearing one thing while someone else in the room had a completely different memory of what was said. Working internationally, where multiple languages are common, can compound this communication problem. I have found only one effective solution to assure that all participants are on the same page. After every meeting, I send each attendee notes from the meeting in their native language. In your personal life, you should also send a message via social media or text confirming exactly where and when your meeting will be. Always include the address in such a communication. For example, in Beijing there are two Ritz Carltons which can be confusing and problematic. Taking a few extra minutes to initiate such contact will make you much more efficient.

53. Create distinct imagery of where you want to be in the future, both personally and professionally. For success on *The Baby Food Diet*, envision the way you will look once you are thin. Browse through a few clothing

第十三章
100个绝佳的建议

的问题,忘了所想象内容的细节。这种情况始于我小时候。曾经一度,一位聪明的朋友告诉我,在床边放一本便笺和一支铅笔,以便在我做了启发性的梦醒来之后使用。现在,当我外出时,会用手机记录下任何新的想法。每到周末,我都会回顾笔记和录音,再也不会忘记下一个价值10亿美元的想法或策略。

52. 在事业和个人生活中,"旁观者清"可能是一个令人困惑的陈腐说法。在开会的时候,我曾经历过,自己听到的是一件事,而与会中的另一个人听到的却完全是另一码事。涉及国际上的工作,使用多种语言是很常见的,这可能会加剧这种沟通问题。我发现只有一种有效的解决办法,可以确保所有参与者能够同步一致。每次会议结束后,我都会使用与会者的母语分别向他们发送会议记录。在个人生活中,你也应该通过社交媒体或短信来确认具体的会面地点和时间,在这样的沟通中一定要包括地址。例如,在北京有两家丽兹卡尔顿酒店(Rit. Carltons),很可能让人因分不清而感到困惑。多花几分钟进行这种联络会让你更有效率。

53. 无论是在个人生活方面还是在职业方面,都需要

stores and admire outfits that you will soon be wearing. Notice a thin, fit person and imagine that is you. The same is true in business, visualize the success you will soon become. Picture your new car, house or fancy office. The beauty of visualization is that it helps in creating and expanding your passion.

54. The first step is crucial when acting on your envisioned ideas. Similar to a red traffic light turning green, it is time to immediately move forward. Start *The Baby Food Diet* today. Do not postpone one day if you really want all the things in your visualizations to become reality.

55. Once you start something, do not give up. Stay the course until you realize success and then rejoice. Likewise, stick with *The Baby Food Diet* because you will soon be celebrating the new you. Don't let periodic failures deter you from reaching your boundless potential.

56. An interesting aspect of human nature regards the occasional and uncomfortable emotions associated with doubt and "buyer's remorse". When people get married, even though they do not tell anyone, some doubts usually exist. The couple might want a child, but feel they must wait until the perfect time. Or the student who wants to apply for a PhD, but assumes the preferred school cannot be afforded. Or the business person who is too scared to expand operations by reasoning it is not the right time. The truth is: it will never feel like the perfect

为将来想要达到的目标创造显著的形象。要想成功坚持**婴儿食品节食法**，可以想象一下自己瘦身之后的样子。去逛几家服装店，欣赏一下你即将穿着的服装。留意一个苗条、健康的人，想象成你自己。同样的道理也适用于生意场，想象自己很快就会成功，畅想一下自己的新车、房子或华丽的办公室。形象化的美妙之处在于其有助于创造和扩展你的激情。

54. 当你按照预期的想法行动时，第一步是至关重要的。就像交通中的红灯变绿灯一样，是时候立即往前走了，今天就开始**婴儿食品节食法**。如果真想让自己想象中的所有事情都成为现实，就一天都不要推迟。

55. 一旦你开始做某事，就不要放弃。要坚持到底，直至成功，然后为之欢呼。同样，要坚持**婴儿食品节食法**，因为你很快就会为全新的自己庆祝。不要让周期性的失败阻止你发挥无限的潜力。

56. 人性中一个有趣的方面与犹疑和"买完即后悔"相关，让人偶尔产生不舒服的情绪。当人们结婚时，即使他们没有告诉任何人，通常还是存在一些疑惑。这对夫

time to do anything. You just need to commit and go for it. If you are doing what you love, at least you are headed in the right direction. Commit to *The Baby Food Diet*, and you will see it is the perfect time.

57. How about implementing the phrase, "I want to do this", to your daily life? We often hear definitive statements like, "I must finish this work", or "I need to write another report before I do anything". By empowering you to make your own decisions, the task at hand becomes easier and probably more enjoyable. For example, when I was in college, I used to go to the library to study to more sharply focus my mind, and it worked. If I stayed in my dorm, students would stop by and ask me to play tennis, go to the gym, or attend a party. Even though I had the willpower to say no, I would start thinking about these fun, alternative options and become inefficient in my studies. Once I arrived at the library, I might have a distractive thought here and there, but would then observe the other diligent students around me and become inspired to immediately return to studying.

第十三章
100 个绝佳的建议

妇可能想要一个孩子,但感觉必须等到最合适的时机。或者是那些想要申请博士学位的学生,会认为首选学校自己负担不起。或者是害怕扩大业务的生意人,会认为时机不对。实际上:永远没有做任何事情的最佳时机,你只需要许下承诺并放手去做。如果你所做的是自己喜欢的事情,至少你在朝着正确的方向前进。去**尝试婴儿食品节食法**,你会发现这是一个完美的时机。

57. 能把"我想做此事"这句话落实到你的日常生活中吗?我们经常听到诸如"我必须完成这项工作"或"我需要在做任何事情之前再写一份报告"之类的明确声明。通过由自己做决定,手头的任务就会变得更容易,也可能更有趣。例如,在我上大学的时候,经常去图书馆学习,以便更加集中注意力,这很有效。如果我待在宿舍里,同学们会顺道过来邀请我打网球、去健身房或参加派对。即使我有意志力说不,也会开始考虑这些有趣的、可替代的选择,使学习变得低效。一旦到了图书馆,我可能也会不时有一些分散注意力的想法,但随后会观察周围其他勤奋的同学,并受到启发,立即恢复学习。

58. Add accountability to your life. Research has proven that we are more likely to achieve something if we disclose our intentions to other people. If you commit to something publicly, the chances for success shoot up because people regularly let themselves down. However, that same person may be reluctant to let *others* down. This applies to *The Baby Food Diet*. Share your intentions with someone who will help monitor your progress from the beginning. That informed person can intermittently check up on you and provide inspiration to forge ahead and attain your goals. This added step of accountability could be the defining difference between failure and success.

59. Preparation for success is imperative to any endeavor. By reading this book, you are preparing yourself for *The Baby Food Diet*. However, reading the book is just the start as there are foods and supplies to buy, daily commitments to be made, and other arrangements that need to be in place to assure a smooth launch. The same is true with a business, you must take the time to prepare for it.

60. Prepare in advance for problems because they will inevitably happen and never ignore them. Complications that go unattended never go away and will possibly get worse. A small, very manageable issue could morph into a very large difficulty if not dealt with immediately. Move quickly because another problem is

第十三章
100个绝佳的建议

58. 给你的生活增添责任感。研究证明,如果我们把自己的意图告诉他人,会更有可能取得成功。如果你公开承诺某件事,成功的机会就会大增,因为人们经常会让自己失望,然而,同一个人可能不愿意让别人失望。这也适用于**婴儿食品节食法**。与那些从一开始就能监督你进步的人分享你的意图。这个了解情况的人可以间歇地检查你的情况,并为你稳步前进和实现目标提供灵感。这一额外的责任感可能是失败和成功之间的决定性区别。

59. 尽一切努力做好成功的准备是必不可少的。通过阅读此书,你正在为**婴儿食品节食法**做准备。然而,阅读本书只是一个开始,因为需要购买食物和用品,需要每天做出承诺,和其他需要到位的安排,以确保顺利实施。做生意也是一样,你必须花时间去准备。

60. 要提前准备,未雨绸缪,因为各种问题将不可避免地发生,永远不要忽视它们。并发症如果不加以治疗永远都不会消失,而且可能会恶化。一个小的、非常容易处理的问题,如果不立即处理,可能会变成一个非常大的困难。迅速行动,因为另一个问题可能即

likely around the corner.

61. For one month track your daily activities. There exist apps you can use to assist in completing this task. Once a month, review where and how your time has been spent. It will reveal if there is adequate balance in your life. You may have gained a few pounds and by monitoring your daily activity (or lack of), you can readjust your daily routines to be in better balance. Then either start or get back on *The Baby Food Diet* before your weight is out of control and adversely affects your life's overall balance.

62. Consistency is a major factor in producing results from *The Baby Food Diet*. Success is not always about being the best, but often about consistency. Constantly working hard will always pay dividends. Stay with *The Baby Food Diet* and remember, it is the forever diet, which is the very definition of being consistent.

63. Earlier, I stressed being an unswerving listener to complement your conversational skills. Watching, hearing, and feeling the emotions of the people you are talking to is an ability well worth learning. Be patient when answering questions, even if the same question has been asked multiple times, and listen very carefully each time. Assuming you are well prepared and know the subject of conversation, you must still listen closely to succinctly address any issues.

第十三章
100个绝佳的建议

将出现。

61. 用一个月的时间跟踪自己的日常活动。有一些应用程序可以帮助你完成这项任务。每个月查看一次自己的时间都是怎样花费以及花在哪些方面，它将揭示你的生活是否足够平衡。你可能已经长了几磅肉，通过监测你的日常活动（或缺乏），你可以调整日常习惯，以达到更好的平衡。然后，在体重失控并对生活整体平衡产生不利影响之前，开始或重新使用**婴儿食品节食法**。

62. 坚持是**婴儿食品节食法**产生效果的一个主要因素。成功并不总与成为最好有关，而是与坚持不懈有关。不断努力，总会有回报。坚持**婴儿食品节食法**，记住，这是永远的节食方法，这正是坚持不懈的定义。

63. 在之前，我强调要做一个坚定不移的倾听者来作为谈话技巧的补充。观察、倾听和感受与你交谈者的情绪是一种非常值得学习的能力。回答问题时要有耐心，即使同一个问题被问过很多遍，每次都要仔细听。就算你准备充分，并且知道谈话的主题，仍然必须仔细倾听，并简洁地解答各种问题。

64. Dress for success is an old adage that has plenty of merit. If you wear clothes that command respect, like a tailored suit, then you are likely to be admired. When starting *The Baby Food Diet*, I advise that you also start dressing differently than in the past. This will jolt your subconscious and send an internal message that changes are in order. In addition to new attire, take the time to clean your fingernails and maintain well-groomed hair. People will take notice of you in a different light. You will also be aware of your evolving new "look" and find it easier to continue advancing on *The Baby Food Diet*. Dressing for success is not just to impress others; it is a reminder to yourself that you are undergoing a life changing physical and mental transformation.

65. If you want to try a revealing test, go into a discussion with open eyes and a blank mind, as if you know nothing at all. This humble approach will force your heart and mind to consider *all* views. To know nothing is a pleasant tactic that makes the other person feel intelligent and special. They will want to teach you everything they know. I mean really, we cannot possibly know everything about the world around us, so why not try this little experiment and see what you learn.

66. The goal of *The Baby Food Diet* is to create an original and lasting habit that will keep you perpetually trim and fit. Of course, it takes time for anything to become habit forming. I have heard new habits can develop in as

第十三章
100个绝佳的建议

64. 人靠衣装是一句老话,真的会有很多好处。如果你穿的衣服能赢得别人的尊重,比如西装,那么你很可能会受到别人的赞赏。当你开始**婴儿食品节食法**时,我建议你也开始穿得和过去有所不同。这将撼动你的潜意识,并发出一个内部信息,有序的改变已经开始。除了新衣服,花点时间清洁指甲,并保持发型整洁。人们会以不同的眼光来注意你。你也会意识到自身改进的新"外观",并发现更容易继续推进**婴儿食品节食法**。注重打扮不仅仅是为了给别人留下深刻印象,也是对自身的一种暗示,告诉自己正在经历一个改变身心的人生转变。

65. 如果你想做一个有启发意义的测试,那就睁开眼睛,清空大脑去参加讨论,就好像你什么都不懂。这种自我谦虚的方法将迫使你的心灵和思想顾及所有的方面。什么都不懂是一种令人愉快的策略,能让对方感到其自身的聪明和特别,他们会将所知道的一切都教给你。我的意思是,我们不可能明晰周围世界的一切,所以为什么不试试这个小试验,看看你能学到什么。

66. **婴儿食品节食法**的目标是建立一个原生的并能持

short as 7 days, but for me it takes much longer, maybe a full month or more. The key is to dive in and keep plowing ahead with the new routine. Once a habit is formed, it no longer takes mental effort to drive your desired action; it becomes second nature.

67. Build your imagination to create even greater success. I have given you the entire template for *The Baby Food Diet*. If you are following the Baby Food Fast or the Baby Food Reduction Diet, there is no place for imagination, as these phases of the diet are clearly outlined. However, once you are on the Baby Food Maintenance Diet, I advise that you use your creative power and develop original healthy meals. Recall, on the Baby Food Maintenance Diet, you are only eating baby food as the substitute for one meal each day. Beyond that, why not use your imagination and calorie counter to create innovative recipes? In the business world, most inventions are due to someone's imagination. Ideas that seem farfetched often become reality. Your culinary concoctions could one day be consumed around the world.

68. The Golden Rule has survived the test of time: "Do unto others as you would like others to do unto you." However, if you really want lasting success, you might also consider this twist on that famous rule: "Do unto others as they would like done unto themselves." The fact is, everybody is different, so different things make them

第十三章
100个绝佳的建议

久的习惯,让你的身体永远保持整洁匀称。当然,任何事情要想养成习惯都需要时间。我听说新的习惯可以在短短7天内养成,但对我来说,会长一些,可能是一个月甚至更长。关键是要投入其中,并按照新的规律不断前进。一旦形成一种习惯,就不再需要精神上的努力来驱动你的行为,它就变成了第二天性。

67. 培养你的想象力,以创造更大的成功。我已经给了你**婴儿食品节食法**的整个模板。如果你正处于婴儿食品禁食期或婴儿食品减食期,没有什么想象的空间,因为这些阶段的饮食是被明确规定了的。然而,一旦到了婴儿食品维持期,我建议你利用自己的创造力开发原生的健康饮食。回想一下,在婴儿食品维持期,你每天只吃一餐婴儿食品作为替代,余下两餐,为什么不用你的想象力和卡路里计数器开发出新的食谱呢?在商界,大多数发明都是源于某人的想象力,看似遥不可及的想法往往会变成现实。你烹饪的食品有一天可能会被全世界追捧。

68. "己所不欲,勿施于人"这条黄金法则经受住了时间的考验。然而,如果你真的想要持久的成功,也可以考虑这条著名法则的变形:"即使己所欲,亦勿施于

happy. They may not prefer being treated the same as you treat yourself. As an example, I like flying economy rather than spending extra money on a higher class of service. To me, flying first class is not more prestigious, and I am very comfortable in coach. Please understand that I fly multiple 12-hour flights every month, so this is a tangible statement. Not surprisingly, the majority of my jet set friends enjoy first class. Instead of urging them to fly coach with me, they are free to enjoy traveling in the style they appreciate. Treating people the way they prefer to be treated will garner you more acclaim than assuming everyone wants to be treated the same as you.

69. Stop making excuses for past failures or for postponing future endeavors. Spending time fabricating excuses disinterests most people seeking success. The sooner you rid the bad habit of justifying your dubious actions, the sooner you will thrive. Stop saying the word "but" whenever possible, as this word is nothing more than an excuse. Instead, put the past behind you and start fresh with a new perspective. I personally have more failure than success stories. One day, I observed that my anecdotes about "almost" successes or terrible failures turned people off. Conversely, when I spoke about my absolute successes, the same people where unanimously interested. Sure, we learn from failures; all experiences can shed light on future achievements. No one

第十三章
100个绝佳的建议

人。"事实上，每个人都各不相同，所以让他们快乐的事情也有所不同。你喜欢怎样对待自己，不见得别人愿意被如此对待。举个例子，坐飞机时我喜欢坐经济舱，而不是花额外的钱买更高等级的服务。对我来说，坐头等舱并不比坐经济舱更有声望，而且我在经济舱里很舒服。请理解我每个月都要飞多次12个小时的航班，所以这是一个有切实说服力的声明。毫不奇怪，我的大多数朋友都喜欢坐头等舱。我会让他们自由地享受自己喜欢的旅行方式，而不是说服他们和我一起坐经济舱。用人们喜欢的方式对待他们，会比认定别人都想和你一样被对待，能得到更多的赞扬。

69. 不要为过去的失败或为推迟未来的努力而找借口。花时间编造借口对大多数寻求成功的人来说没有意义。你越早改掉为自己可疑行为辩护的坏习惯，就会越快成长。尽可能不要说"但是"，因为这个词只不过是一个借口而已。相反，请把过去抛诸脑后，用新的视角重新开始。就我个人而言，失败的故事比成功的故事要多。某天，我发现我那些关于"几乎"成功或非常失败的轶事让人们感到厌烦。相反，当我谈到绝对成功时，同样这些人一致表示对此感兴趣。然而，我们能从失败中学习，所有的经

wants to hear irrelevant stories about how many times you failed on previous diets. However, when you talk about your weight-loss accomplishments on *The Baby Food Diet*, a captive audience will be listening because success sells.

70. You must believe in miracles. What this really means is you must believe in yourself. The mere fact that you are alive and existing on this earth is a miracle. You have been given the most wonderful gift ever and that is being you, so cherish every day. Do not take this incredible opportunity for granted. With that in mind, accept and appreciate who you are and do your very best 24/7. And that all starts with bettering your health. *The Baby Food Diet* will achieve results today and forever.

71. Never blame others for your failures. To cast blame is the same as making excuses and not moving ahead. If your business has failed, then start another one. If you fault everyone else, you are not taking responsibility. This must change, and it must change now. If you are overweight, there is no one to blame but yourself for poor eating habits. You were the one who gorged on the unhealthy food that was placed in front of you. You were the slacker who refused to exercise. Accept the culpability, put it behind you, forgive yourself, and strive forward with a new-found outlook. And it all begins with *The Baby Food Diet* which will give you the confidence and self-esteem needed to be a new person. You

第十三章
100 个绝佳的建议

验都能对未来的成就有所启发。没有人愿意听一些无关的故事，比如你以前减肥失败了多少次。然而，当你谈论在**婴儿食品节食法**方面的成就时，听众就会被吸引，因为成功是有卖点的。

70. 你必须相信奇迹。这意味着你必须相信自己。你能活着并存在于这个世界上，这一事实本身就是一个奇迹。你已经被赐予了最美妙的礼物，那就是你自己，所以请珍惜每一天。不要把这个不可思议的机会视为理所当然。记住这一点，接受并欣赏你自己，每天、时时刻刻都要做到最好。而这一切都始于改善自身健康。**婴儿食品节食法**从现在到未来都将取得成果。

71. 永远不要把自己的失败归咎于别人。推卸责任就像找借口而不前进一样。如果你的买卖倒闭了，那就再开一家；如果你指责别人，就是没有担当。这必须改变，而且现在就改。如果你超重了，不能怪别人，只能怪自己不好的饮食习惯。你是那个贪吃摆在面前的不健康食物的人，你是那个拒绝锻炼的懒鬼。接受自己的过错，并抛诸脑后，原谅自己，用一种全新的视角奋力前行。这一切都从**婴儿食品节食法**开始，它会给你成为一个崭新的人所需要的自信

will no longer blame others, but instead proudly talk openly about your success.

72. People do have a sixth sense. As such, I believe in listening to my intuition. Invariably, when I operate instinctively, I win. When I don't, I lose. It's that simple for me. Success is not achieved by taking one day and one win at a time. It is a cumulative process of accumulating many wins over a longer period of time. Your goal should be to become successful and stay successful. *The Baby Food Diet* is perfect for this model, as you first lose weight, then keep lean with the maintenance diet. Listen to your intuition and enjoy ongoing and never-ending success.

73. Accept that not everything can be changed. Never worry about what cannot be altered. Instead concentrate on what can be changed and make that happen. One thing you can change is how much you weigh by developing proper eating habits and following *The Baby Food Diet*. We encounter obstacles every day. Some are easy to overcome. Others are not, for any number of reasons. The point is that obstacles overcome will shine a bright light on you and your integrity. For example, I had a very important meeting in Beijing where there was a huge convention of foreign leaders, making it nearly impossible to hire a taxi. Concerned with being tardy, I kept calm and brain-stormed a solution. First, I located a share bike and rode to a subway station. Then took the

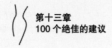

和自尊。不要再责备别人,而是自豪地公开谈论自己的成功。

72. 人们确实有第六感。因此,我相信并听从自己的直觉。一直以来,只要我凭直觉行事,就一定会赢。反之,则会输。对我来说就是这么简单。成功不是通过一天中的一次胜利而获得的,而是一个在较长时间内汇集许多成功经验的累积过程。你的目标应该是取得成功并保持成功。**婴儿食品节食法**是这一案例的完美模型,因为你首先需要减肥,然后依靠维持节食保持体形。听从你的直觉,享受持续不断的成功。

73. 要接受不是所有事物都能够改变。永远不要担心无法改变的事情;相反,应该集中精力去做可以改变的事情,并让改变发生。你可以改变的一件事就是自身的体重,通过养成正确的饮食习惯和遵循**婴儿食品节食法**就能改变。我们每天都会遇到障碍,有些是容易克服的,另外一些则很难,原因有很多,重要的是,你所克服的障碍将照亮你和你的为人。例如,我在北京参加了一个非常重要的会议,而此时的北京正举行一个有很多外国领导人参加的国际会议,使得这个时段几乎不可能叫到出租车。由

train and disembarked closer to my destination. Finally, I mounted a second share bike and arrived at the meeting just in time. The wealthy gentlemen at the gathering would never consider taking a bike or the subway for that matter. When I told them what I did to arrive punctually, they showed me added respect because my integrity proved how important they are to me.

74. Stay active and productive. Doing nothing is boring and creates inefficiencies. Have you ever been to a very busy restaurant and were amazed by the swift waiters and expeditious service? Then you go to another, less crowded restaurant where the service is dreadful. The waiters are standing around chit-chatting with each other, and the food arrives slower and cold. What distinguishes the two eateries? People with nothing to do are not efficient. Spare time is ideal for self-improvement and should be utilized by exercising, joining a class, or reciting out loud your Self-Talk affirmations. If you are on *The Baby Food Diet*, also use any extra time to create delicious meals that are the right portion within the appropriate calorie range.

第十三章
100个绝佳的建议

于担心迟到,我保持冷静,绞尽脑汁想出了一个解决办法。首先,我找到了一辆共享单车,骑到一个地铁站,然后乘坐地铁,在离目的地最近的一站下了车,最后,我又骑上另一辆共享单车及时赶到了会场。参加聚会的富人们从来不会因为这件事而考虑骑自行车或乘地铁,当我告诉他们我是如何准时到会之时,他们对我更加尊重,因为我诚恳的为人证明了他们在我心中的重要性。

74. 要保持活跃和高效。什么事情都不做会很无聊,而且会造成效率低下。你是否去过一家非常热闹的餐馆,那里麻利的服务员和快捷的服务让你感到十分惊讶?然后你去另一家人少的餐馆,那里的服务很差。服务员们站在旁边互相聊天,饭菜端上来的时候又慢又冷。这两家餐馆有什么不同?无所事事的人工作效率不会高。业余时间是自我完善的理想时间,应该通过锻炼、参加课程或大声朗读自己的自言自语而对其加以利用。如果你正在尝试**婴儿食品节食法**,也要利用任何额外的时间来制作美味的食物,当然,分量要合理并控制在适当的卡路里范围内。

75. Chart out your long-term goals in separate plans: one-year, five-years, and ten-years. Use them as guidelines to your future. Such objectives are not meant to be set in stone. The purpose is to plot a meaningful direction and keep you from floundering. Obviously, things do change because the world is dynamic, not static, and so your strategy should be flexible. Once per year, I advise a reassessment of your plans to adjust to your more current desires and goals. Within those yearly plans, you must reference staying on *The Baby Food Diet*, your catalyst for enduring physical and mental success.

76. In life, giving and receiving produce varying responses among individuals. Some people gain more satisfaction from giving to others rather than receiving or taking from others. Some believe that giving is good karma and will eventually result in getting back what you gave and perhaps even more so. There are very few guarantees in life, and the idea of cause and effect is not always accurate. However, *The Baby Food Diet* is a prime example of a cause and effect guarantee. If you follow it precisely, you will definitely be thinner. If you never give seeds the water and soil they need, you will never receive a crop to harvest. Give *The Baby Food Diet* your total dedication and commitment, and you will receive the terrific results you could only dream of in the past.

第十三章
100个绝佳的建议

75. 把你的长期目标分开规划:一年计划、五年计划和十年计划,用它们来指导你的未来。这些目标并不是一成不变的。这样做的目的是为了描绘一个有意义的方向,让你不至于陷入困境。显然,事情会发生变化,因为世界是动态的而并非静态的,所以你的策略也应该是灵活的。我建议每年重新评估一次你的计划,以适应当前的愿望和目标。在这些年度计划中,你必须提到**婴儿食品节食法**——这一持久的身心成功的催化剂。

76. 在生活中,给予和接受在个体之间产生不同的反应。有些人从给予别人中获得更多的满足感,而不是从别人那里得到或索取。这些人认为给予是善业,最终自己的付出会有收获,甚至更多。生活中几乎没有什么保证,因果关系的概念并不总是准确的。然而,**婴儿食品节食法**是一个能有因果保证的最好的例子。如果你严格按照它执行,你一定会变得更瘦。如果你从来不给种子所需的水和土壤,就永远不会有收成。全身心投入**婴儿食品节食法**,你会得到过去只能梦想得到的最棒的结果。

77. We've all been told: "Never waste time." I do not agree and believe that everyone needs to take breaks where they have absolutely no responsibility. Such a pause can allow your mind to wander and discover unexpected creativity. It is not wasting time to be alone for a while in our thoughts or to occasionally lay in bed for an extra hour. I think such relaxation helps rejuvenate you by recharging your soul. Now and then, give yourself permission to watch television and chill out. Conversely, the flagrant wasting of endless days lying in bed or watching television will eventually destroy you. Again, a balance needs to be implemented to make you happier, healthier and ultimately more productive.

78. Patience is a compulsory trait of the successful person. Impatience is one of the main reasons many leaders are under so much pressure and fail. Of course, leaders should push their team to perform, but within reason. Sitting around waiting, instead of being proactive, is not being patient. It is being lazy. There are times for the boss to push for results, and other times to simply wait for results. The person who only pushes is out of balance and will fail. The same is true of *The Baby Food Diet*. When you begin with the Baby Food Fast, be patient because this first phase of the diet will take time to pay dividends. Lack of patience can cause anxiety, frustration, and impel you to stray from the goal of losing weight. Remember the old saying: "Patience is a Virtue."

第十三章
100个绝佳的建议

77. 我们都曾被教导:"永远不要浪费时间。"我不同意这一观点,我认为每个人都应该在完全没有责任的时候休息一下。这样的停顿可以让你的大脑漫游,发现意想不到的创造力。独自一人自行思考一段时间,或者偶尔在床上多躺一个小时,都不是浪费时间。我认为这样的放松可以让你的灵魂重新充满活力。偶尔,要允许自己看电视放松一下。相反,公然无休止地躺在床上或看电视,最终会毁了你。再说一次,需要实现一种平衡,让你更快乐、更健康,最终更有效率。

78. 耐心是成功人士的必备品质。缺乏耐心是许多领导人承受很大压力及导致失败的主要原因之一。当然,领导者应该督促他们的团队去执行,但要在合理范围内。坐以待毙,而不是积极主动,不属于有耐心,而是懒惰。有些时候需要老板推动才能有结果,另一些时候只是等待结果。光会推动的人失去了平衡,也会失败。**婴儿食品节食法**也是如此。当你开始婴儿食品禁食期时,要有耐心,因为节食的第一阶段需要一些时间才能有回报。缺乏耐心会导致焦虑、沮丧,并迫使你远离减肥的目标。记住那句老话:"耐心是一种美德。"

79. Always focus on solutions, not problems. Problems are everywhere and easy to identify. Often a company will have a method for their employees to address pertinent issues which are crucial to improving operations. A well-run company will expect all employees who detect a problem to also provide a resolution. However, a recent study revealed that for every 20 identified problems identified by employees, only three workers also offered solutions to those issues. More specifically, you are reading this book due to your overweight problem and are seeking a viable solution. *The Baby Food Diet* delivers the very best solution: all the pragmatic answers to realizing a thinner body and the accompanying tranquility are answered by this diet. Winning people find and implement winning solutions to problems.

80. If you want something, ask for it. Don't be shy or you shall be denied. It all starts with enthusiasm for something desired, whether it's money, love, respect, a job, or the loss of excess weight. Enthusiasm is a powerful emotion that can quickly grow into passion. Before starting *The Baby Food Diet*, you may still have doubts. Now that you've read and learned more about its effectiveness, you'll become more enthusiastic about starting immediately. After about six weeks on the program, you'll see fascinating results, and that enthusiasm will turn to passion which will keep you devoted to *The*

第十三章
100 个绝佳的建议

79. 要关注解决方法，而不是问题本身。问题无处不在，很容易发现。通常，公司会为员工提供一种解决相关问题的办法，这对改善运营至关重要。一个经营良好的公司会期望所有发现问题的员工也能提供解决方案。然而，最近的一项研究表明，每20名发现问题的员工中，只有3名员工提出了解决这些问题的方案。更具体地说，你读这本书是因为自身超重问题，并正在寻找一个可行的解决方案。**婴儿食品节食法**提供了最好的解决方案：所有实用的答案都通过这一节食方法给出，用以实现一个更瘦的身体和随之而来的宁静。成功的人对于问题会找到并实施成功的解决方案。

80. 如果你渴望什么，就去寻求。不要害羞，否则你会被拒绝。一切都始于对渴望之物的热情，无论是金钱、爱情、尊重、工作还是减肥。热情是一种强大的情感，它能迅速成长为激情。在开始**婴儿食品节食法**之前，你可能还心存疑问。现在，经过阅读并了解更多关于其有效性的知识后，你会对于立即开始更加热情。大约六周后，你会看到迷人的结果，这种热情会转化为激情，能让你永远专注于**婴儿食品节食法**。热情是前进的动力，是追求自己渴望之物的动力。

Baby Food Diet forever. Enthusiasm is the driving force that keeps you going, keeps you asking for what you desire.

81. Right now is the best time to take action. I keep stressing that to be successful, you must stop procrastinating. Realize that time does indeed fly, so don't waste one more second and spring into action today. Nothing from yesterday matters. Today, you are free to make new choices in your life and act upon them. As you walk down the street, notice the small nuances like rustling leaves on the trees or the chirping birds because they will change from day to day. You, too, change daily. Commit to *The Baby Food Diet* today and start seeing positive changes every day and forever.

82. Being 99 percent committed to something is not enough. If you really want success, you must be completely dedicated. In business, the 100 percent rule must be observed by the company founders for success to be attained. They will face a multitude of challenges and any commitment less than 100 percent will likely lead to failure. Of course, in the actual running of a business, the 100 percent rule is not actually feasible due to unforeseen variables. On a personal level like dieting, you must be 100 percent committed to *The Baby Food Diet* to be successful. Call it a leap of faith, but you must fully commit to reach the target weight you have dreamed of your whole life.

83. People rarely stop smoking, no matter what

第十三章
100个绝佳的建议

81. 现在是采取行动的最佳时机。我一直强调,要想成功,必须停止拖延。你要认识到时间确实会飞逝,所以不要再浪费一秒钟,今天就行动起来,昨天发生的一切都无关紧要。今天,你可以随便在生活中做出新的选择并付诸行动。当你走在街上的时候,注意那些细微的差别,比如树叶的沙沙声和鸟儿的啁啾声,因为它们每天都在变化。你也一样,每天都在改变。从今天开始投身**婴儿食品节食法**吧,从此每天都能看到积极的变化,直到永远。

82. 做事只投入百分之九十九是不够的。如果真心想要成功,你必须全身心投入。在商界,要想取得成功,公司创始人必须投入百分之百的努力。他们会面临诸多挑战,任何低于百分之百的承诺都可能导致失败。当然,在企业的实际运营中,由于无法预见的变量,百分之百规则实际上是不可行的。在个人层面上,如节食,你必须百分之百投身于**婴儿食品节食法**才能成功。你可以将之称为信心的飞跃,但你必须全身心投入到实现毕生梦想的目标体重中去。

83. 人们很少能戒烟,不管他们怎么说,直到健康出

they say, until health concerns force them to quit. Others make
a conscious decision to drop the nasty habit because they simply
want to for whatever reason. The same is true of drug addicts and
alcoholics. They must want to change before any meaningful progress
can be made. You are about to start a diet and must sincerely want
to see change to make it all the way through. This book has given
you many tools to create the mind-over-matter engine that drives
this process. This means it is solely up to you to determine its
success. Look at yourself in the mirror and say three times in
a loud voice: "I have made a choice to be thin and healthy
and to commit 100 percent to starting *The Baby Food
Diet* today."

84. Taking the approach that "Life is a game" is
a good philosophy. Most of us like competition, and
everyone loves to win; it is human nature. So why not
play the game by taking one day at a time, with the
goal of being victorious on all levels of life. You'll
win some and lose some. This approach will make
living more enjoyable and entertaining. *The Baby
Food Diet* can be the game in which you will become
the grand champion.

85. Expectations have much to do with future
results. If you expect to be successful, your chances are
exponentially improved. Of course, expectations demand
effort. There is a direct correlation between the height

问题才会迫使他们戒掉。有些人有意识地决定戒掉这一坏习惯,因为不管出于什么原因,他们就是想戒掉。吸毒者和酗酒者也是如此。在取得任何有意义的进展之前,他们必须试图改变。你即将开始节食,必须真心实意想要看到改变,这样才能坚持到底。本书为你提供了许多工具来创造驱动"精神胜于物质"这一过程的引擎,这意味着成功完全取决于你。看着镜子里的自己,大声地说三遍:"我已经做出了一个选择,我要瘦身和健康,今天就开始百分之百地投身于**婴儿食品节食法**吧。"

84. 采取"生活就是一场游戏"这一方法很有哲理。我们大多数人喜欢竞争,而且每个人都喜欢赢,这是人类的天性。所以,为什么不玩一天一次的游戏,目标是在生活的各个层面都取得胜利。你会有赢有输,这种方法将使生活更加愉快和有趣。**婴儿食品节食法**可以是使你成为冠军的游戏。

85. 人的期望与未来的结果有很大关系。如果期望成功,你的机会就会成倍地增加。当然,期望需要努力。期望值越高,需要付出的努力就越多。因此,认真写下自身的目标,就是在加强期望,并为未来创造一个

of the expectations and the energy required to achieve that goal. Therefore, by writing down your goals, you are reinforcing and creating a clear vision for the future. Expect to lose weight, put in the effort, and you surely will succeed.

86. Life should be fun. Fun is our reward for working hard. Make eating baby food fun, and it will be easier to stay with the diet. Send social media photos of yourself devouring baby food. Maybe wear a T-shirt that says: "I Eat Baby Food." The more fun you make something, the easier it is to stay with it.

87. When you are at work, stop living for the weekend. Every day is special and should be appreciated. Sure, there are bad days and good days, but to exist only for the weekend can also be depressing. Make new friends at work, find something, anything you like about the job, and you will better enjoy each day. My point is not to allow yourself to be miserable five out of seven days a week by desperately awaiting weekends to only appreciate life. *The Baby Food Diet* will change your despondent attitude by instilling more self-confidence for every moment of every day.

88. Always have faith that success will indeed arrive. Believe in hope as it will open many doors. Then seize the opportunity and take action with the support of faith, and you will know success.

清晰的愿景。期待减肥,付出努力,你一定会成功。

86. 生活应该是有趣的。乐趣是我们努力工作的回报。要让吃婴儿食品变得有趣,这样就更容易坚持下去。在社交媒体上发你自己吞食婴儿食品的照片,或者穿一件写着"我吃婴儿食品"的t恤。你做的事情越有趣,就越容易坚持下去。

87. 当你工作的时候,要在周末停下来好好享受。每一天都是特别的,都应该珍惜。当然,有好日子也有坏日子,但只在周末享受生活也会令人沮丧。在同事中结交新朋友,在工作中找到兴趣点,你就会更好地享受每一天。我的意思是,不要让自己一周七天中有五天因为绝望地只能等待周末享受生活而痛苦不堪。**婴儿食品节食法**将通过时时刻刻不断地灌输更多的自信,来改变你沮丧的态度。

88. 永远相信成功一定会到来。相信希望,因为它能打开许多扇门。然后抓住机会,在信念的支持下采取行动,你就会懂得成功。

89. Baby steps grow into long, meaningful strides, just like *The Baby Food Diet*, and eventually reach their intended destination, premeditated success. No step is too small, as every inch forward is one closer to achieving your goals. For example, by weighing yourself once per month while on *The Baby Food Diet,* you will always see accomplishment each time you step on the scale. Even the smallest improvement is meaningful and ample reason to keep moving forward. Soon all these little things will add together to create the success you are seeking with *The Baby Food Diet*.

90. I make it a habit to compliment other people to elevate my mood. It makes me feel good to notice positive things about other people. That's right, I praise other people more for *my* mood than theirs! If you take the time to find something good about everyone you meet, it will restore faith in yourself and humanity. I only give true and sincere compliments. Sometimes for certain people, it can be difficult to find something to compliment. However, if you look close enough, there is always something special about everyone. Maybe it is something as small as a tie clip, a brooch, the sound of their voice or the color of their eyes. Whatever it is, this practice will start the new relationship on a positive note. As a result of this practice, it will also stimulate positive feelings about you in the eyes of the other person. As odd as this may sound, the very compliment you give will also help with your diet. The logic

第十三章
100个绝佳的建议

89. 婴儿的每一小步都会成长为有意义的一大步，最终达到他们的目标。这种谋划好的成功，**就像婴儿食品节食法**一样，没有哪一步是太小的，因为每向前走一寸就离目标更近了一步。例如，当你遵循**婴儿食品节食法**时，每月称一次体重，每当你走上体重秤时，总是会看到成就，即使是最小的改进也是有意义的，并且是继续前进的充足理由。很快，所有这些小进步加在一起，就会通过**婴儿食品节食法**，创造出你所追求的成功。

90. 我养成了一种习惯，通过称赞别人来提升自己的情绪。注意到别人的积极方面让我感觉很好。没错，我赞美别人主要是为了我自己的情绪，而不是为他人！如果你花时间去发现遇到之人的优点，会使你更有自信心和人情味。我只给予真诚的赞美。有时候，对某些人来说，很难找到值得称赞的东西。然而，如果你仔细观察，会发现每个人都有一些特别之处。也许是像领带夹、胸针以及他们说话的声音或眼睛的颜色这样的小细节，不管是什么，这种做法都会让你们的关系有一个积极的开端。这样做的结果，也会激发别人对你的好感。虽然这听起来很奇怪，但你给出的赞美也会对自身的节食有所帮助。道理很简单：如果你自我感觉良好，就会产生自

is simple: if you feel good about yourself, self-esteem is created and it is much easier to continue with the diet.

91. Plan to be happy. It sounds crazy and so simple, but why not make it a goal? The truth is, just trying to be happy will never materialize on its own. Happiness is the result of activities and achievements. So, if your goal is to be joyful, then you need to experience accomplishments. It all starts with creating the new you and setting realistic and achievable goals that will bring you contentment. *The Baby Food Diet* is the first and best route to your future happiness. From the success you realize with the diet, you can add more and more positive things into your life, and before long you will be happy every day. It is true that one good deed does lead to another.

92. "Curiosity may have killed the cat", as the old saying goes. A nice response to this saying is: "Curiosity also brought it back." What this means is that being constantly curious can help you uncover things that will make you a more complete person. If you are inquisitive about certain subjects, then read books, take courses, surf the internet, observe firsthand. Maybe your inquiries will lead to a new job, career, love life. To be curious means having an insatiable appetite to learn. Are you curious about what it will take to become thin? The answer: start *The Baby Food Diet* today.

93. If a really golden opportunity knocks on your

第十三章
100个绝佳的建议

尊感,继续节食就会容易得多。

91. 要计划快乐。这听起来很疯狂,其实也很简单,但为什么不把它作为一个目标呢?事实上,仅仅想着要快乐是永远不会实现的。幸福是活动和成就的结果。所以,如果你的目标是快乐,那么你需要经历成就。这一切都始于创造一个全新的你,设定现实可行的目标,让自己感到满足。**婴儿食品节食法**是你通往未来幸福的第一个,也是最好的途径。在从节食中实现成功的基础上,你可以添加越来越多积极的东西到生活中,不久你就会快乐每一天。的确,一件好事会导致另一件好事。

92. 俗话说:"好奇害死猫。"对这句话的一个很好的回应是:"好奇也能救活猫。"这意味着,保持好奇心可以帮助你发现能让自己人生更完整的方法。如果你对某些学科感到好奇,那就读书、上课、上网、亲自观察。也许询问一下,就会给你带来一份新的工作、新的职业、新的爱情生活。好奇意味着永不满足的学习欲望。你想知道怎样才能变瘦吗?答案是:从今天就开始**婴儿食品节食法**。

93. 如果一个真正的绝好机会来敲门,你必须抓住它,

door, you must seize it or you might not get it again. By waiting to start *The Baby Food Diet*, you are burning valuable time and missing an excellent chance to experience positive changes and a better attitude, so why wait one more second? The diet will shed unwanted pounds and serve as a catalyst that benefits every single aspect of your life. Opportunities only come to those who are prepared. Start the diet today and relish all the ancillary advantages that stretch far beyond your weight and health.

94. Learn to stretch once a day to stay young and limber. As you get thinner, you will become more active and take up various forms of exercise. Being agile means less chance of injury. Ten minutes of stretching each day will go a long way toward keeping you healthy. If you dislike stretching on your own, then consider joining a yoga class, a martial arts class, or possibly a dance class. After you start *The Baby Food Diet*, you will have more energy and actually *want* to work out!

95. Frustration is a terrible, inescapable emotion to experience at any time. However, continued frustration over changeable elements in your life should be addressed and resolved. For a while, my phone battery was running out of power at significant times of the day, and the back-up battery I carried was not enough. It was maddening. The obvious fix was to carry multiple extra batteries which required that I also tote an extra bag. The

否则你可能不会再得到。坐等观望**婴儿食品节食法**，是在浪费宝贵的时间，并错过一个体验积极变化和更好状态的绝佳机会，那么为什么还要再等一秒钟呢？节食可以减掉多余的体重，并成为一种催化剂，让生活的方方面面都受益。机会只给有准备的人。从今天开始节食，享受那些远远超出体重和健康之外的附加好处。

94. 学会每天做一次伸展运动，以保持年轻和柔韧。当你瘦身后，会变得更活跃，可以进行各种形式的锻炼。灵活意味着受伤的机会更少。每天 10 分钟的伸展运动对保持健康大有裨益。如果你不喜欢独自做伸展运动，可以考虑参加瑜伽课、武术课，或者也可能是舞蹈课。一旦你开始**婴儿食品节食法**，会有更多的能量，并且真的想要运动！

95. 挫折感是一种在任何时候都无法逃避的可怕情绪。然而，生活中不断变化的因素所带来的挫折感应该得到解决。有一段时间，我的手机总是在一天中最重要的时候没电，我带的备用电池也不够用。这让人很抓狂。最简单的解决办法是多带几块电池，这就需要我多带一个包。即使有更多的负担，权衡起来也是值得的。我之所以提到这一点，是因为当你开始

trade-off was worth it regardless of the burden of more baggage. I mention this because as you start *The Baby Food Diet*, you might also be required to carry an extra bag for the baby food and protein. For women it is easier, as most already carry a purse, and baby food does not take up much space. The benefits of not missing meals is worth the hassle of carrying an extra bag. The good news is that losing weight is not frustrating and instead exhilarating.

96. One hundred years ago, the average person died at approximately 48 years of age. Many reasons contributed to this fact, including decaying teeth. I suggest as part of your weight loss program, regularly check and clean your teeth. Healthy teeth do not limit your intake of nuts, fruits, and other nutritional foods, unlike dentures or no teeth at all. Plus, pearly-white teeth boost your self-esteem and inspire you to smile more brightly. Why not visit a qualified dentist and get a maintenance check-up for your own good?

97. I am often asked about drinking other beverages beside water while on the diet. In the beginning, you can drink black coffee or straight tea. Do not add sugar, artificial sweeteners or dairy products. Once you are on the Baby Food Maintenance Diet, you can expand what you drink and even add sugar in small dosages to your coffee or tea.

98. Genuine respect for others is crucial in life. The more you respect others, the more they will respect you.

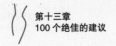
第十三章
100个绝佳的建议

婴儿食品节食法时,你可能也需要额外带一个包来装婴儿食品和蛋白质。对于女性来说,这更容易些,因为她们大多数本身就会带个包,而且婴儿食品也不会占太多空间。不用少吃一餐的好处是值得多带一个包的。好消息是,减肥并不令人沮丧,反而令人振奋。

96. 100年前,人的平均寿命大约是48岁。造成这一事实的原因有很多,包括牙齿问题。我建议,作为减肥计划的一部分,要定期检查和清洁你的牙齿。不同于假牙或没牙,健康的牙齿不会限制进食坚果、水果和其他营养食物。此外,洁白的牙齿能增强你的自信,让你笑得更灿烂。为了自己的利益,为什么不去看一个合格的牙医并做一次维护检查呢?

97. 我经常被问到,在节食的时候,除了喝水是否能喝其他饮料。刚开始的时候,你可以喝黑咖啡或清茶,不要添加糖、人造甜味剂或乳制品。一旦到了婴儿食品维持期,你可以扩大饮料范围,甚至可以在咖啡或茶里添加小剂量的糖。

98. 在生活中,真心尊重他人至关重要。你越是尊重别人,别人就越是尊重你。有些人天生刻薄,最好不要理

Some people are just downright mean, and it is best to ignore those kinds of cynics. I talked about complimenting people earlier, but showing respect is every bit as important as a compliment. Also, by respecting others, you are respecting yourself.

99. Always live in the present. Focus and pay attention when in a meeting. Show respect by not taking or making phone calls. Even looking at your phone during a meeting sends a message that you are bored or disinterested in what is being discussed. Instead, pay attention and take notes to show that you care about the topics and speaker. When it is your turn to talk, offer questions and comments that are intelligent and pertinent to what was presented and the subject at hand. As a side note, you will find that people you have met in the past but not seen for some period of time, will notice that you are losing weight. If they ask about it, you can respond with enthusiasm on the merits of *The Baby Food Diet*, and how it is positively changing your life. Keep it short and on point because many people only ask to be polite and do not really care. If they continue to ask questions, then go into more detail. You will be amazed at how eagerly and naturally that you expound on the positive attributes of *The Baby Food Diet* in such a situation.

100. Let me conclude my suggestions with a summary of the aforementioned practical thoughts. Treat yourself with kindness. Prioritize what matters most in life.

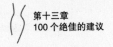

会那些愤世嫉俗之人。我之前说过,要赞美别人,但是对他人表示尊重和赞美一样重要。而且,你尊重他人,就是在尊重自己。

99. 永远活在当下。开会时要集中注意力,不要接电话或打电话,以示尊重。即使是在开会的时候看手机,也会让人感觉你对正在讨论的内容感到无聊或无趣。相反,应该集中注意力并做笔记,以表明你很在意所讲的主题和演讲者。当轮到你发言的时候,提出一些明智的问题和评论,并与所谈论的内容和当前主题相关。另外,你会发现,以前见过但有一段时间没见的人,会注意到你正在减肥。如果他们问起来,你可以热情地介绍**婴儿食品节食法**的优点,及其如何积极地改变了你的生活。回答要简短并切中要点,因为很多人只是出于礼貌而问,并不真正在意。如果他们继续询问,那就回答得更详细些。你会惊奇地发现,在这种情况下,自己是多么急切而自然地阐述**婴儿食品节食法**的积极特性。

100. 让我以对上述实际方法的总结来结束我的建议。要善待自己。优先考虑生活中最重要的事。找一个支持小组。正常睡眠。每天喝 8 杯水。在节食的时候写日

CHAPTER 13
100 Amazing Suggestions

Find a support group. Get proper sleep. Drink 8 glasses of water each day. Keep a journal while on the diet and maybe forever. Do not compare yourself to others. Say "no" when it is time to say "no". Step out of your comfort zone and safely try something new. Stay in contact with family and friends. Join exercise groups. Appreciate and celebrate the wonderful you. Take up a hobby. Learn from mistakes. Be organized and punctual. Be a great communicator. Be responsible and accept blame. Keep calm and do not lose your temper. Think before you speak. Dream big and often. Be a pleasant person. Never get embarrassed, but be humble. Forgive others and move on. Apologize when you are wrong. Be a happy, sincere person. Listen to music to change your mood. Believe in yourself. Remember that it is healthy and okay to say no. Listen to your instincts. Learn new things. Be patient. Ask for what you want. Be fully committed. Clean your house and throw away any extra clutter. Be thankful and appreciate the person that you are. Ask for help when you need it. Find a mentor. Be a mentor. Never harp on problems unless you have a solution. Smile often. Surround yourself with winners. Practice gratitude daily. Compliment others. Dress for success. Maintain daily personal hygiene. Have a good memory. Notice small things. Last of all and most importantly, start *The Baby Food Diet* today!

第十三章
100 个绝佳的建议

志,也许永远都要写。不要拿自己和别人比较。该说"不"的时候要说"不"。走出舒适区,尝试新事物。与家人和朋友保持联系。参加健身小组。欣赏和赞美最棒的自己。培养一项爱好。从错误中学习。做事有条理并守时。做一个最好的沟通者。承担责任,接受责备。保持冷静,不发脾气。三思而后行。要时常怀有远大的梦想。做一个令人愉快的人。永远不要感到尴尬,但要谦虚。原谅别人,继续前进。做错事以后要道歉。做一个快乐、真诚的人。听音乐来改变心情。相信自己。记住,说"不"是有益无害的。听从自己的直觉。学习新事物。要有耐心。想要什么就去寻求。要全身心投入。打扫房子,扔掉多余的杂物。感激并欣赏你自己。当需要帮助的时候要寻求帮助。找一位导师。成为一名导师。除非你有解决办法,否则不要喋喋不休地谈论问题。经常微笑。常和成功人士在一起。每天练习感恩。赞美他人。注重穿着。保持日常个人卫生。要有好记性。关注小事。最后,也是最重要的,从今天就开始**婴儿食品节食法**吧!

CHAPTER 14

**Short Summary of the
Baby Food Diet**

第十四章

婴儿食品节食法概要

CHAPTER 14
Short Summary of the Baby Food Diet

Here is a short summary of the different phases of *The Baby Food Diet:*

I. Baby Food Fast (not to last more than 14 days) :
- Eight glasses of water everyday
- One glass of water before each meal (3 meals)
- One glass of water after each meal (3 meals)
- Two glasses of water before bed
- You can drink more than eight glasses of water each day, but eight is the minimum

BREAKFAST: Protein shake made with water (glass of water before and after)

LUNCH: One baby food for lunch (glass of water before and after)

DINNER: One baby food for dinner (glass of water before and after)

II. Baby Food Reduction Diet: (normal 6 weeks to 180 days)
- Eight glasses of water everyday
- One glass of water before each meal (3 meals)
- One glass of water after each meal (3 meals)
- Two glasses of water before bed
- You can drink more than eight glasses of water each day, but eight is the minimum

第十四章
婴儿食品节食法概要

以下是对**婴儿食品节食法**不同阶段的简要总结:

I. 婴儿食品禁食期(最多不能超过14天):

- 每天喝8杯水
- 餐前喝1杯水(共3餐)
- 餐后喝1杯水(共3餐)
- 睡前喝2杯水
- 你可以每天喝水超过8杯,但8杯是最低标准

早餐:由水搅拌的蛋白奶昔(餐前餐后各饮1杯水)

午餐:1份婴儿食品作为午餐(餐前餐后各饮1杯水)

晚餐:1份婴儿食品作为晚餐(餐前餐后各饮1杯水)

II. 婴儿食品减食期(通常6周~180天):

- 每天喝8杯水
- 餐前喝1杯水(共3餐)
- 餐后喝1杯水(共3餐)
- 睡前喝2杯水
- 你可以每天喝水超过8杯,但8杯是最低标准

早餐:蛋白奶昔和1份婴儿食品作为早餐

BREAKFAST: Protein Shake for breakfast and one baby food

LUNCH or DINNER: One baby food (lunch and dinner are interchangeable)

LUNCH or DINNER: One meal as follows:

- Some type of fish or Chicken
- One piece of toast without butter
- Vegetables to your liking but a small portion
- Can substitute one banana for the vegetables
- One salad with only vinegar dressing
- Possibly one more baby food

III. Baby Food Maintenance Diet: (forever)

- Eight glasses of water everyday
- One glass of water before each meal (3 meals)
- One glass of water after each meal (3 meals)
- Two glass of water before bed
- You can drink more than eight glasses of water each day, but eight is the minimum

BREAKFAST: One protein shake and one baby food (or substitute eggs for protein shake)

LUNCH: One baby food

DINNER: Anything you want if you are following the rules of portion control

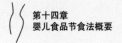

第十四章 婴儿食品节食法概要

午餐或晚餐：1份婴儿食品（午餐和晚餐是可以互换的）

午餐或晚餐：以下任意一餐

- 1份鱼肉或鸡肉
- 1片吐司不加黄油
- 1小份喜欢的蔬菜
- 可以将蔬菜替代为1根香蕉
- 1份只浇醋汁的沙拉
  - 可能再加1份婴儿食品

III. 婴儿食品维持期：（永久）

- 每天喝8杯水
- 餐前喝1杯水（共3餐）
- 餐后喝1杯水（共3餐）
- 睡前喝2杯水
- 你可以每天喝水超过8杯，但8杯是最低标准

早餐：1份蛋白奶昔和1份婴儿食品（或用鸡蛋代替蛋白奶昔）

午餐：1份婴儿食品

晚餐：只要你遵循分量控制的规则，可以吃任何想吃的食物

IV. Weight Monitoring and Weight Fluctuations in the Maintenance Phase:

- Weigh yourself once a month
- Remember, The Baby Food Diet is based on portion control, so during this phase you can experiment with enjoying small portions of your favorite foods to satisfy cravings, such as two Hershey kisses or 5 potato chips
- Keep a food journal
- Base your proportions during the Baby Food Maintenance Diet on your desired weight and keep within 5 pounds of your goal
- If you go more than 5 pounds over your goal weight, switch from the Baby Food Maintenance Diet to the Baby Food Reduction Diet or the Baby Food Fast.
- Don't beat yourself up if you occasionally have a cheat day from the Baby Food Diet. Just don't let too many pounds creep on before you return to healthy eating patterns based on The Baby Food Diet.

V. Other Optional Mind/Body Enhancements to *The Baby Food Diet*:

- Self-Talk
- Visualization of a lean and healthy body

第十四章
婴儿食品节食法概要

IV. **维持阶段的体重监控及波动：**

- 每月称 1 次体重
- 要记住，**婴儿食品节食法**是基于分量控制，所以在此阶段你可以尝试享受一小份喜爱的食物以满足欲望，如 2 块好时巧克力或 5 片薯片
- 记食品日志
- 在婴儿食品维持期，以理想体重为基础，保持自身体重不要超过目标体重 5 磅
  - 如果超过目标体重 5 磅以上，从婴儿食品维持期转到婴儿食品减食期或婴儿食品禁食期
    - 在坚持**婴儿食品节食法**期间，如果你偶尔有一天作弊，不要自责。只希望在你回到以**婴儿食品节食法**为基础的健康饮食模式之前，不要长出太多的赘肉

V. **其他可选的精神 / 身体方面对于婴儿食品节食法的增强功能：**

- 自言自语
- 想象苗条而健康的身体
- 以积极的心态和方式对待生活
- 平衡
- 冥想、深呼吸，或其他放松技巧

- A positive mental attitude and approach towards life
- Balance
- Meditation, deep breathing, or other relaxation techniques
- Exercising for its physical and mental benefits, but not necessarily weight loss
- Multivitamins
- Other nutritional supplements for general health

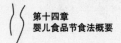
第十四章
婴儿食品节食法概要

- 锻炼对身心有益，但并非减肥必需
- 复合维生素
- 其他整体健康所需的营养补充剂

ACKNOWLEDGMENTS

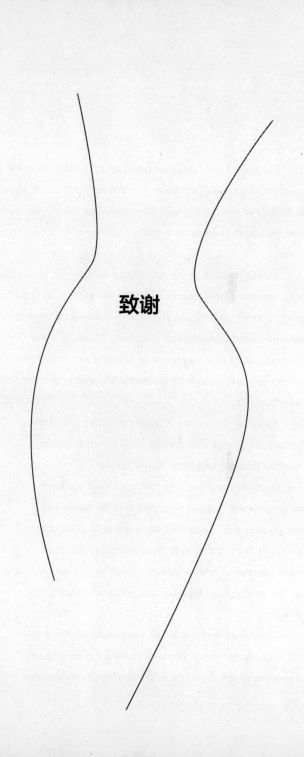

致谢

ACKNOWLEDGMENTS

The Baby Food Diet is not based on a scientific study, but on common sense and observation. For that reason, I asked a couple of friends to observe and add ideas, including Tony Villa. Tony is a super businessman, song writer, producer and best of all a great friend.

One suggestion he offered: "Since most people are on the go, why not also have prepacked protein similar to baby food that's easy to carry and less messy." Tony continually contributes new ideas and is committed to making *The Baby Food Diet* the best weight loss program ever. Many of the menu items on our app and website (www.thebabyfooddiet.com) are from Tony.

Tony has never had a weight problem, but does understand human nature and knows which foods are beneficial and detrimental to our respective bodies. For example, the greatest tennis coach does not have to be a superb tennis player to know the game and the greatest golf coach does not have to be a scratch golfer to know the strategy. All successful educators must possess an intense knowledge of human nature to become highly respected and proficient mentors, like Tony.

Tony does have one health issue worth mentioning. He has high cholesterol, like his beloved father who died as a result at an early age. You must be aware of your personal

致谢

婴儿食品节食法并非基于科学研究,而是基于常识和观察。出于这个原因,我邀请了几位朋友进行观察并补充了一些想法,其中包括托尼·维拉(Tony Villa)。托尼是一位成功的商人、歌曲作者、制作人,是我最好的朋友。

他提出的一个建议是:"既然大多数人都很忙碌,为什么不准备一些与婴儿食品类似的蛋白质,以方便携带,减少麻烦。"托尼不断贡献出新的想法,并致力于使**婴儿食品节食法**成为一直以来最好的减肥方法。我们应用程序和网站(www.thebabyfooddiet.com)上的许多菜单选项都来自托尼。

托尼从来没有超重问题,但他了解人类天性,知道哪些食物对各种类型的身体有益、哪些有害。例如,最棒的网球教练不一定非得自己是优秀的网球运动员,才明白怎么打比赛;最棒的高尔夫教练也不一定非得自己是零差点球手,才懂得击球策略,等等。所有成功的教育者都必须具备丰富的人类本性知识,才能成为像托尼这样受人尊敬和专业精通的导师。

托尼确实有一个健康问题值得在此一提。他胆固醇高,就像他深爱的父亲一样,其父因此很早就去世了。因

hereditary in order to increase the chances of a long and healthy life. Tony takes medications to control his cholesterol, and his doctors tell him that he will live to be more than 100 years old. For this reason, he is also concerned about the foods he consumes. Tony does everything possible to prolong a healthier life, which means eating responsibly and exercising every day. *The Baby Food Diet* is the solution he has always sought to help fight high cholesterol. However, if you are on prescribed medications for blood pressure or cholesterol issues, never stop taking your pills just because you are thin. These medications should only be stopped if a doctor tells you to do so. Many hereditary problems cannot be contained by diet alone, and it is best to do everything available to promote an extended, healthier life. That means taking your prescriptions and enjoying a healthy diet.

Tony is so passionate about *The Baby Food Diet* that he has dedicated his life to spreading the word about the many benefits it has brought to him and his family. Tony contributes recipes, tips to keeping fit, and psychological strategy to stay on the diet. *The Baby Food Diet* and his daily medication shave removed the stress that he felt due to his looming cholesterol complications. Tony is a very positive person who adds great value on our journey to a happy mental and physical life.

Mr. Benjamin Lye is another friend who has been

致谢

此你一定要了解自身家族遗传病史,以便增加长寿和健康生活的机会。托尼服用药物来控制胆固醇,医生告知他可以活到 100 岁以上。由于这个原因,他也关心自己所吃的食物。托尼尽一切可能延长其健康生活,这意味着要谨慎饮食和每天锻炼。**婴儿食品节食法**是他一直寻求能帮助其对抗胆固醇问题的解决方案。然而,如果你正在服用治疗高血压或胆固醇问题的处方药,千万不要因为自己很瘦就停止服药。只有当医生告诉你可以停止服用时,你才能停。许多遗传问题不能仅靠饮食来控制,最好能够尽一切可能延长健康的生活,这就意味着既要吃处方药,也要享受健康的饮食。

托尼对**婴儿食品节食法**充满了热情,并全身心投入到传播婴儿食品节食法给他自己和家人带来的诸多好处中。托尼帮忙提供食谱、保持健康的秘诀,以及能够坚持节食的心理策略。**婴儿食品节食法**和日常药物已经消除了他由于担心胆固醇并发症所带来的压力。托尼是一个非常积极乐观的人,他为我们的身心生活之旅增添了巨大的价值。

本杰明·莱先生(Benjamin Lye)是另一位帮助完善节食法的朋友,但他的故事与托尼的大不相同。本(本杰明)一

helpful in perfecting the diet, but his story is much different from Tony's. Ben has always been overweight and never had any success at any level on dieting. While Ben is not yet thin, he has dropped 60 kilo's, and the weight is still falling off. Ben has had so much success that I asked him to help edit the book. Ben is an enthusiastic person full of life and fun. He is an Australian and loves life more than anyone I know.

Ben strictly follows *The Baby Food Diet* and also exercises daily at the gym to reach his goals. He has more energy now than ever before, which has changed his mental approach to life as well. Even the way he dresses has improved. *The Baby Food Diet* has altered the way Ben looks at himself. Now, he is a person with no limits and aiming for greatness at every level.

These are only two of the many people who have significantly benefited from *The Baby Food Diet*. Why not be the next person to benefit from *The Baby Food Diet*? As I have emphasized throughout this book, *The Baby Food Diet* is the permanent solution to your weight problems. More than that, it will impact all areas of your life in a positive way.

I asked another good friend, Victoria Li, to proof read what I had written and focus on spelling and grammar. Victoria was my highly educated personal assistant for years and is a very sweet person. She has a

致谢

直超重,此前在节食方面从来没有取得任何成功。虽然本目前还没完全瘦下来,但他已经减掉了60公斤,而且体重还在下降。本已经取得了很大的成功,所以我请他帮忙编辑此书。本是一个充满活力和乐趣的热心人,他是澳大利亚人,比我认识的任何人都更加热爱生活。

本严格遵循**婴儿食品节食法**,并且每天去健身房锻炼,以达到自身目标。他现在比以往任何时候都精力充沛,这也改变了他对待生活的态度,甚至他的穿着也有了改善,**婴儿食品节食法**改变了本看待自己的方式。现在,他已不再受限,其目标是在每一个层面上都取得伟大的成就。

这只是许多从**婴儿食品节食法**中大大受益者中的两个人。你为什么不成为下一个从**婴儿食品节食法**中受益的人呢?正如我在本书中一直所强调的,**婴儿食品节食法**是解决体重问题的永久方案。不仅如此,它还会以积极的方式影响你生活的方方面面。

我还请了另一位好朋友维多利亚·李(Victoria Li)来校对我写的东西,重点是拼写和语法。维多利亚受过高等教育,曾是我多年的私人助理,非常可爱。她对细节

ACKNOWLEDGMENTS

keen eye for detail and did find and correct many errors. For this, I am grateful to her.

However, the real story is not the edits she made, but instead, her genuine reactions. Victoria is thin, has always been thin, and will forever be thin. She was so captivated with the book's beneficial content that she now tells everyone she knows or meets to become a part of *The Baby Food Diet* revolution. In fact, every day Victoria submits new ideas and recipes to make the diet even better.

As this book ends, let me remind you when it comes to dieting, there are no fresh ideas. The difference between all diets is often the approach, philosophy, and discipline. The strategy of eating baby food to lose weight is not new. Other diets with baby food as a primary meal source are only designed for quick body fat loss. *The Baby Food Diet* as detailed here, is a comprehensive weight management program that will not only rapidly shed pounds, but also keep your weight under control forever. Commit to *The Baby Food Diet* today and enjoy a new, happier, slimmer you!

To receive updates and the latest diet news, recipes, and information please regularly check out www.thebabyfooddiet.com or our app. Share your success with others in your community, so they can also be thin, healthy, and beautiful -- like YOU!!

致谢

有敏锐的观察力,确实发现并改正了许多错误。为此,我很感谢她。

然而,现实故事并不是她所做的编辑工作,而是她真实的反应。维多利亚很清瘦,一直如此,而且会长此下去。她被书中有益的内容所吸引,现在她告诉她所认识或遇到的每一个人,使之成为**婴儿食品节食法**革命的一分子。事实上,维多利亚每天都会提交新的想法和食谱,让节食法变得更好。

在本书结束之际,请让我提醒你,说到节食,没有什么新鲜的方法,所有节食法的区别通常是方式、哲理和纪律。吃婴儿食品减肥的策略并不是新方法。其他以婴儿食品为主要食物来源的节食法,只是为了快速减肥而设计的。这里详细介绍的**婴儿食品节食法**,是一个全面的体重管理计划,不仅能快速减肥,还能让你的体重永远可控。今天就开始**婴儿食品节食法**吧,享受一个全新的、快乐的、苗条的你!

要想获得最新的节食新闻、食谱和信息,请定期访问 www.thebabyfooddiet.com 或我们的应用程序,在群组里跟其他人分享你的成功,这样他们也可以像你一样苗条、健康、美丽!!!

BONUS CHAPTER

A Few More Thoughts

附录

再说几点

Congratulations and thank you for reading this amazing book *The Baby Food Diet*. I hope you will begin the journey immediately to create a new wonderful you. I suggest in one month to re-read the book. I have a few last thoughts that you may find interesting.

Earning credibility and creating momentum are always part of the process of reaching success on any venture. New ideas at first will face challenges. In the beginning, the process is always slow and difficult. However, with a step by step approach, success will be realized. Often this success will be much faster than originally thought.

At the start, many people will make fun of "The Baby Food Diet", thinking that it is a fad or a joke. However, before long this will start to change. Many stories of success will be posted on social media. People will comment on social media sites how much weight they have lost. Often, a person might add to their post on how their health has improved. Other people will point out how their energy levels have increased since starting the diet. Some will talk of how the diet has empowered them and that their self-esteem has dramatically increased.

However, it's much more than just social media posts that will propel the diet to success. For example, people will find that they personally know someone, a

附录
再说几点

诚挚恭喜并感谢您读完神奇的《婴儿食品节食法》，希望您会迫不及待地踏上下一段精彩旅程，成为更好的自己。建议您能在1个月后再来复习一遍这本书。最后，我还有几点想法，或许您有兴趣听一听。

在任何一段冒险旅程中，提升信任感并获取源源不断的动力是成功抵达彼岸的必要条件。任何新概念在最初总会面临重重质疑。一开始，前进的步伐看起来步履维艰、困难重重，但只要循序渐进，一定会一步步走向成功。而且这份成功通常还会让你有些措手不及。

最初，很多人会嘲笑**婴儿食品节食法**，认为它不过是个噱头或者笑话。但不用太久，人们的看法就会大为改观，社交媒体上会陆续出现很多成功案例，人们也会迫不及待地在这些网站上评论自己成功减重了多少，很有可能有人会在帖子中表明自己的健康状况有所提升，有些人会指出自开始节食以来他们的精力如何增加，还有人会评论这种节食方法是如何大大地提高了他们的自尊心。

然而，这种节食方法广泛流行的助推剂绝不仅仅是社

friend or acquaintance who is now slim, healthy and happy due to the baby food diet.

You may notice a person wearing a shirt that proudly says, "I look great because I am on *The Baby Food Diet*". With each step, more and more momentum and credibility will be created. This momentum will never stop, as this diet is the permanent solution to weight management and better health. With *The Baby Food Diet*, once the weight is off, it will stay off. A myriad of health problems will disappear.

All at once the baby food diet goes from a joke to a respected lifestyle choice. Instead of a person being laughed at for being on the baby food diet, they will be admired. Unlike fad diets every year more and more people will join the baby food diet evolution and revolution.

The Baby Food Diet's popularity will grow forever and never wane. You will find medical professionals, trainers, nutritionists and others endorsing *The Baby Food Diet* lifestyle. Yes, over the next 10 to 20 years, it is reasonable to think that half the world population will follow *The Baby Food Diet*. The diet is logical, sensible, healthy, easy to follow and best of all it works. Welcome to *The Baby Food Diet* revolution and evolution and a new you.

附录
再说几点

交媒体的帖子。比如,有人会发现自己身边的朋友、熟人在用婴儿食品节食法之后变得纤细苗条、健康快乐。

你也可能会看到一个身穿写着"I look great because I am on The Baby Food Diet"(我的美丽全靠**婴儿食品节食法**)这些大字的T恤。每向前一步,信任感与动力都会逐步递增,它永远不会枯竭,因为这种节食方法是管理体重和迈向健康的永恒法宝。一旦通过**婴儿食品节食法**减重成功便永不反弹,而无数健康问题也会迎刃而解。

霎时间,**婴儿食品节食法**会从一个被人唾弃的笑话变成受人尊敬的生活方式。那些原本对**婴儿食品节食法**嗤之以鼻的人也会对它肃然起敬。**婴儿食品节食法**不会像其他网红节食法一样昙花一现,每年都会有大批志愿者加入它的行列。

人们对**婴儿食品节食法**的推崇只会与日俱增。医疗专家、运动教练、营养学家等人也都会为**婴儿食品节食法**背书。我们有理由相信,在未来的10至20年间,世界上会有半数人口采用**婴儿食品节食法**。因为它逻辑清晰、合理实用、健康方便,更因为它切实有效。

BONUS CHAPTER
A Few More Thoughts

It is clear from blogs, articles, news stories, the stock market and other media sources that eating habits are radically changing. As an example, plant-based meat substitutes are in the news on a regular basis. With more access to information, people are beginning to think, before they consume food products. People now understand that while calories do matter, other factors must also be weighed. One lesson I learned on Wall Street is to always follow the trend. This is the reason: many food giants are in the process of developing plant based alternative foods.

Fruits and vegetables grown organically without pesticides and chemicals are now in every grocery store in America. The price for such items is higher than on non-organic choices, but they are the biggest sellers. This is a trend which cannot be ignored. The fact is more and more people every day are choosing the healthy alternative. The trend will continue to expand and have a multiplier effect on the demand.

The real challenge is growing healthy food. It all starts with the soil. This is a big statement, and you must listen. The soil from which a crop is grown is the key element in producing healthy fruits and vegetables. Soil is the building block of the crop. Build a home with flimsy wood and you end up with a house that will eventually fall. You must know if the soil is truly organic, free of pesticides and has not been exposed to man-made

附录
再说几点

欢迎加入**婴儿食品节食法**的革命战队，成为更好的自己。

博客、文章、新闻、股市和其他媒体均表明饮食习惯正发生着翻天覆地的变化。例如，植物性肉类替代品已经成了新闻中的常客。人们获取的信息逐渐增多，在购买食品时也就变得更加挑剔。他们了解卡路里的重要性，但也会通盘考虑其他因素。审时度势是华尔街给我上的重要一课。很多餐饮界的巨擘都在开发植物性替代食品。

现在，你在美国的每家食品超市里都能找到未使用农药和化学制品的有机蔬果。虽然它们比非有机食品贵，但销量却遥遥领先，势头不容小觑。选择健康食品的人日益增多，因此这种势头只会持续走高，对需求量也会产生乘数效应的刺激。

然而，种植健康农作物才是真正的挑战。万物源于土壤。请务必记住我接下来的重要声明。作物生长的土壤决定了蔬果的健康，土壤是作物的本源。正如朽木筑屋难承风雨，你也必须了解土壤本身是否真正有机，无农药及合成化肥的影响。土壤是种植无致癌物质等

fertilizers. Soil is one of the key components to grow healthy non-carcinogenic foods.

Even in geographical regions where the soil appears healthy, it is often flawed by over growing (not resting the land), chemicals such as non-organic fertilizers and pesticides. Such practices have over time have caused much of the worlds soil to become infertile. Even in cases where the soil is fertile, problems may still arise. Often these practices of pesticides and non-organic fertilizers have created diseased soil.

There is great news, this damage can be repaired by a modern style of composting. I am aware of a private company that provides soil solutions. They are soil scientists and have taken the centuries-old art of composting and added elements to the equation. The results are unbelievable. They can cure soil diseases and make infertile soil once again fertile.

You are part of the solution, so please separate your trash. The organic foods you throw away will indirectly contribute to you next healthy meal. Soil is so important that I believe in the future when you buy organic foods, a sign will be displayed as to what type of soil was used to grow that fruit or vegetable.

What you eat will also impact mood swings, energy and lifespan. We are in an era that evokes a higher consciousness on food and life choices. Why not make

附录
再说几点

健康作物的关键。

即便是土壤状况良好的地区，通常也会受到过度耕种（土地休息不充分）和非有机农药化肥的影响。长此以往便导致全球严重的土壤贫瘠问题。肥沃的土壤上也依然问题频发。农药和非有机化肥的使用通常都会导致土壤病态。

但好消息是，现代化的混合堆肥可以修复这类损害。我发现，有家解决土壤问题的私人公司，组织土壤专家研究已有数百年历史的混合堆肥技术，在古老的配方中又加入了几味元素，他们的成果出乎意料，不仅治愈了病态土壤，还让贫瘠的土地焕发了生机。

你也是保护土壤的重要一分子，所以请一定做好垃圾分类。你扔掉的有机食品能间接为你的健康餐食做出贡献。土壤的重要性不言而喻，我甚至相信总有一天，你购买的有机食品上会标明种植这种蔬果的土壤种类。

你的饮食也会对情绪波动、精力和寿命产生影响。当今时代，人们更加注意食品和生活选择。何不尽量简化

your food choices easier?

*The Baby Food Die*t is your solution to quick weight loss, permanent body management and good mental health. A way to get slim, become healthy and maintain your target goals. If you look up *"The Baby Food Diet"* on the internet you will find similar names, but they are much different. What you find on an Internet search is simply a fad diet to lose weight quickly. This is not one of those diets, with the exception that you will lose weight at an extremely rapid pace.

The big difference with "The Baby Food Diet" from others is that in addition to losing weight fast, it will stay off forever. *The Baby Food Diet* is hardly for babies, but instead for adults, who wish to lose weight and keep it off. As a side benefit of *The Baby Food Diet*, don't be surprised if your energy soars. As a result of the diet, your transformation will be nothing short of incredible. You will look great! Your attitude will be positive! If you want to change what is going on around you, then you must first address what is inside you. This means your mental mindset and the food you consume.

The diet will redefine how you view yourself in relation to others. It will cause you to appreciate yourself more than in the past. You will discover that you are looking forward to each new day with enthusiasm. The

附录
再说几点

你的食品选项呢?

婴儿食品节食法能同时解决快速减重、心理健康及长期身材管理的问题,是瘦身、健体、维持目标身材的有效方式。如果在网上搜索**婴儿食品节食法**,你也能找到相似的标题,但内容却相去甚远。网络上搜索到的大多是只追求快速减重的网红节食法,而这种方法除了让你极速减重以外,其他方面都截然不同。

婴儿食品节食法的一大特点就是快速减重之后还能保证永不反弹。**婴儿食品节食法**并不适用于婴儿,而是为希望减重并保持身材的成人而设计。采用这种方法后你也很有可能会感到精力大幅提升,这是它带来的额外奖励。这种节食法还一定会给你不可思议的转变,你的外表会非常棒!你的态度会更乐观积极!如果你想要改变周遭的环境,就必须先调节自己的内在,即你的思想观念和食物选择。

通过这种节食方法,你会重新审视自己与他人的关系,更加欣赏自己,也会发现自己更加热爱生活,对每一天都充满期待。这本书不仅包含饮食方法,它蕴含的

philosophies, not just the way of eating, that are discussed in the book are your answer to creating a slim body and becoming satisfied with your life. The diet is empowering, powerful and life altering. It is a well-thought-out and time-tested solution to weight management. The diet is a catalyst for keeping good mental and physical health.

I hope you will tell all your friends what you have learned as a result of reading this book. As you have realized the book was designed to be a fast and easy read, giving the reader simple steps to follow and motivational tools to create success. The book is in fact "Life Changing". As suggested earlier, I advise that you reread this book in one month. This diet will impact your life in such a positive way that you will be thankful beyond what you can imagine.

People in today's world care about what they consume and their health. Over the last few years the number of vegetarians and vegans has grown substantially. Some people are vegans, as they are against the killing of animals and others for various health reasons. Regardless, being a vegan or vegetarian is no longer a closet type of lifestyle, as it is embraced by a large segment of our population. Vegetarians and vegans tend to be people looking to improve their health — people like you, that is why you have just read *The Baby Food Diet*.

The Baby Food Diet works for vegetarians, vegans

附录
再说几点

哲理更是你开启健康纤体、幸福生活的钥匙，这种节食方法拥有改变你一生的强大力量。这一精心制定的方法经过了时间的考验，是体重管理的有效途径，也是心理和身体健康的双效催化剂。

我希望你能将从这本书中读到的感悟分享给朋友。这本书的设计适合快速、方便地去阅读，为读者的成功提供简单的步骤和激励工具，你也应该对此有所体会。这本书能"改变一生"。如前所述，我建议1个月后你能再复习一遍这本书，这种节食方法会对你的生活产生超乎想象的积极影响，而你也会对它备加感激。

当今世界，人们非常在乎自己的消费选择和身体健康。在过去几年中，素食主义者和严格素食主义者的人数大幅增加。有些人是因为反对杀害动物及其他不同原因而成为严格素食主义者的；无论如何，大部分人都接受了素食主义或严格素食主义的生活方式，所以他们也无须再有所隐藏。素食主义者和严格素食主义者多数都抱有提高身体健康的目标。相信你也一样，所以才会读完这本**婴儿食品节食法**。

and meat eaters. Very small modifications can be made to satisfy all three types of diet. If you are a vegan, you will be pleased. If you are a vegetarian, you will be pleased. If you are a meat eater, you will be pleased.

The Baby Food Diet is the most flexible diet you will find anywhere. Making modifications to the diet is simple and can be done instantly. With life, it is always about choices and with *The Baby Food Diet* you are in control. For example, if you are a vegan, whey protein is not acceptable as it is made from dairy produces. This is not a problem, as you can substitute plant-based proteins such as soy, pea, hemp or others.

The Baby Food Diet works for three primary reasons: 1) It is easy to follow 2) It forms new "bite-sized mini-habits" and 3) Portion control is automatic and super easy.

The first reason the diet works is that it is simple and easy to follow. There is nothing complicated about the diet. This is not only the world's greatest diet; it is also the diet that will put you on autopilot. Since you have already read the book, which probably took you only ninety minutes, along with some shopping for food, you are ready to begin. As an added benefit, you will find that your food costs will go down due to the *The Baby Food Diet*.

The second reason the diet works is that it forms

附录
再说几点

婴儿食品节食法对素食主义者、严格素食主义者和肉食主义者统统适用,只要稍做调整,它就能完全满足三种饮食的不同要求。无论你是严格素食主义者、素食主义者或是肉食主义者,你都会对它非常满意。

无论你在哪里都能使用**婴儿食品节食法**,它是最灵活的饮食方式,调整起来也很简单、快速。生活就是一个接着一个的选择,而选择了**婴儿食品节食法**你就掌控了全局。例如,如果你是严格素食主义者,那就不能选择乳制品来吸收乳清蛋白。但这并不是问题,你可以转而从大豆、豌豆等食品中吸收植物蛋白。

婴儿食品节食法的效果主要源于以下三个特点:1)简单易行;2)它能培养"滴水穿石的习惯";3)自动、简便的分量控制。

这种节食方法效果明显的第一个原因就是它简单易行,完全不复杂。它不仅是全世界最棒的节食方法,还能帮你直接开启"自动驾驶"模式,顺利踏上饮食调整的道路。你应该仅用了90分钟就读完了这本书,只要再去购买一些食品,你就准备完毕,可以出发了。你

habits. Developing habits is a huge part of *The Baby Food Diet*. A habit is something a person does repeatedly and on a regular basis. Once a habit is formed, it will become an unconscious action that is part of who you are. A habit takes anywhere from a few days to a few months to be formed.

Habits are formed due to the frequent repetition of something along with attached mental imprinting. Once you have a habit it can added to a routine which will further enforce the unconscious regular behavior.

We have all noticed the warning on packages of cigarettes: to be careful as smoking them may be habit forming and bad for your health. Well that's the bad side of a habit. The point of the warning is that once a habit is formed it is hard to change.

We only hear the negative side of habits, such as the cigarette example. How about the positive side of habits? Why not purposely form good habits, things that benefit you since they are so hard to get rid of? *The Baby Food Diet* uses this idea of creating good habits as a gigantic factor in creating success.

附录
再说几点

还会发现它的另一个好处,就是你的食品开销会大大降低。

第二个原因就是它会培养你的习惯。养成习惯是**婴儿食品节食法**的重要一环。习惯是指一个人定期重复做的事情,习惯一旦形成就会成为一种无意识的行为,变成你生活的一部分。然而,要养成一种习惯则需要几天到几个月时间不等。

习惯的形成是因为经常重复某件事并形成精神印记。习惯养成后就会添加到你的常规日程中,从而进一步强化无意识的重复行为。

我们都注意到了香烟包装上的警示,提示人们吸烟可能会养成有损健康的习惯。其实这就是习惯的负面效果。这个警告也告诉我们习惯一旦形成就难以改变。

我们只听到了吸烟这类习惯的消极影响,习惯的积极影响又有哪些呢?既然习惯很难改掉,那为何不有意识地养成良好习惯,做对你有益的事?**婴儿食品节食法**就是利用养成良好习惯以获得成功的重要方法。

Developing permanent good habits which will help you get slim and help you stay there forever is a major element of the diet. *The Baby Food Diet* will become a new permanent good habit. Now that you have read the book, I am sure that you will agree that *The Baby Food Diet* is a brilliant approach and is easy to follow.

The third reason the diet works is the portion control that it creates. Portion control is often listed as a key component in many popular diets. The problem with portion control is that often it is too complicated to implement. Nobody wants to carry measuring cups or scales everywhere they go. Most meals on *The Baby Food Diet* come in pouches that are premeasured, making them easy to consume.

Portion control is automatically created with *The Baby Food Diet*. Better than this, the meals can easily be carried in your pocket or bag. There are diets which have advance-prepared meals, but they are normally bulky and often require refrigeration. With *The Baby Food Diet* no refrigeration is required. Most of the meals on *The Baby Food Diet* are premeasured portions, making them easy to carry and consume.

When you put it all together, *The Baby Food Diet* is your answer to a new better you. A simple way to become the person you want to see when you look in the mirror. Best of all is that the mental benefits of the diet are "Life Changing". The book teaches you more than a way to eat, it

附录
再说几点

永久养成利于纤体、保持身材的良好习惯是这种方法的重点。**婴儿食品节食法**将会成为伴你一生的好习惯。读完这本书,我相信你也能够认同**婴儿食品节食法**方便易行、切实有效。

而这种节食法有效的第三个原因则是它的分量控制特点。很多流行的节食方法都十分注重分量控制,但它们的一大问题就是过于复杂而难以执行。没有人愿意走到哪里都随身携带量杯和秤。**婴儿食品节食法**中使用的食物大都被预先称重、分成小袋,非常方便食用。

婴儿食品节食法自动解决分量控制的问题:你甚至还可以轻松地把食物放到口袋或书包里,随身携带。其他节食方法也会有预先准备好的食物,但一般都很占地方还需要冷藏。**婴儿食品节食法**的食物则不需冷藏,它使用的食物大多数都预先按分量包装,方便携带也便于食用。

通盘考量之后,你会发现,想要成为更好的自己,**婴儿食品节食法**就是答案。它能轻而易举地让你变得更好,让你满意地看着镜子里的自己。最重要的是,它带给你心理健康的提升足以"改变一生"。这本书不仅教给你一种饮

creates a new you. You will learn a philosophy, which is the key to getting started and staying with the diet.

Indeed, the mental aspects, not just overeating, are why you are overweight in the first place. Overeating is mental and once this is controlled, so is your weight. Getting your head in the right place carries the same importance as eating correctly. This book does more than telling you what to eat. The book teaches the reader mindset techniques that assure success. *The Baby Food Diet* gives you a way to eat that will become a habit — habits that will be turned into routines and help create a new you. *The Baby Food Diet* is the answer you have been looking for. Now it is up to you to begin this incredible journey to become slim, healthy and happy.

The Baby Food Diet is a lifestyle, based on easy-to-consume superfoods. I visualize half the world's population eventually following the diet, as it is a permanent solution to all the wonderful things that life offers. This may be somewhat extreme thinking, but the diet is highly effective. You can always contact us at: thebfdiet.com, thebabyfooddiet.com, info@thebabyfooddiet.com or info@thebfdiet.com. We hope you have enjoyed reading this book as it will change your life for the better and forever.

**附录
再说几点**

食方法，更创造了一个全新的你。你学到的是一种哲学，这才是你开始并坚持一种节食方法的关键。

事实上，超重的原因绝不仅仅是暴饮暴食，更与心理层面相关。暴饮暴食其实就是一种心理问题，一旦控制了它，你也就控制住了体重。找准正确的思想心理状态与保持正确的饮食习惯同等重要。这本书绝不仅仅告诉了你应该选择何种饮食，它还教给你调整心态的技巧，以便你最终达成目标。**婴儿食品节食法**教给你一种饮食习惯，它会转变为你常规日程中的一步，帮助你成为全新的自己。**婴儿食品节食法**是你一直在寻找的答案。现在，就由你自己决定，是否要踏上这段非同凡响的旅程，成为苗条、健康、快乐的自己。

婴儿食品节食法是一种生活方式，利用的都是方便食用的超级食品，它能带来生活中所有的美好。因此我能想象，世界上最终有一半的人口都会使用这种方法。虽然听起来有些天方夜谭，但它的效果绝对毋庸置疑。您可以随时通过以下方式联系我们：thebfdiet. com, thebabyfooddiet.com, info@the-babyfooddiet.com，或者 info@thebfdiet.com。衷心希望您喜欢这本书，相信它会让您的美好生活越来越好。

In addition to the principles laid out in this book, there will be daily updates via our YouTube channel, podcasts, an app, website (www.thebabyfooddiet.com), live events, blogs, and so much more. So, let your journey to weight control success begin by joining *The Baby Food Diet* revolution!

除了本书列出的原理之外,我们还将通过YouTube频道、播客(podcasts)、应用程序、网站(www.thebabyfooddiet.com)、现场活动、博客等方式进行每日更新。所以,让你成功的减肥之旅,从加入**婴儿食品节食法**的革命开始吧!